D1384334

BOWLING GREEN STATE UNIVERSITY
DISCARDED
LIBRARY

PRODUCTION OF HIGH STRENGTH CONCRETE

PRODUCTION OF HIGH STRENGTH CONCRETE

by

M.B. Peterman
R.L. Carrasquillo
The University of Texas at Austin
Center for Transportation Research
Austin, Texas

NOYES PUBLICATIONS
Park Ridge, New Jersey, U.S.A.

Library of Congress Catalog Card Number 85-25924
ISBN: 0-8155-1057-8
Printed in the United States

Published in the United States of America by
Noyes Publications
Mill Road, Park Ridge, New Jersey 07656

10 9 8 7 6 5 4 3 2 1

Library of Congress Cataloging-in-Publication Data

Peterman, M.B.
Production of high strength concrete.

Bibliography: p.
Includes index.

1. High strength concrete. 2. Road materials--
Texas. I. Carrasquillo, R.L. (Ramón Luis), 1953–
II. Title.
TA440.P448 1986 625.8'4 85-25924
ISBN 0-8155-1057-8

Foreword

The criteria for selection of concrete materials and their proportions to produce uniform, economical, high strength concrete are presented in this book. The recommendations provided are based on a study of the interactions among components of plain concrete and its mix proportions, and of their contribution to the compressive strength of high strength concrete. These recommendations will serve as guidelines to practicing engineers in selection of materials and their proportions for producing high strength concrete.

There are definite advantages, both technical and economical, in using high strength concrete in structures today. For example, for a given cross section, prestressed concrete bridge girders can carry greater service loads across longer spans if made using high strength concrete. In addition, cost comparisons have shown that the savings obtained through the use of high strength concrete members are significantly greater than the added cost of the higher quality concrete. This book should prove to be a valuable document for civil engineers and related personnel.

The information in this book is from *Production of High Strength Concrete* prepared by M.B. Peterman and R.L. Carrasquillo of the Center for Transportation Research, Bureau of Engineering Research, the University of Texas at Austin, for the Texas State Department of Highways and Public Transportation, Transportation Planning Division, Austin, Texas, and the U.S. Department of Transportation, Federal Highway Administration, October 1983.

The table of contents is organized in such a way as to serve as a subject index and provides easy access to the information contained in the book.

Advanced composition and production methods developed by Noyes Publications are employed to bring this durably bound book to you in a minimum of time. Special techniques are used to close the gap between "manuscript" and "completed book." In order to keep the price of the book to a reasonable level, it has been partially reproduced by photo-offset directly from the original report and the cost saving passed on to the reader. Due to this method of publishing, certain portions of the book may be less legible than desired.

ACKNOWLEDGMENTS

This work was sponsored by the Texas State Department of Highways and Public Transportation and the Federal Highway Administration, and administered by the Center for Transportation Research at The University of Texas at Austin. The authors would like to acknowledge the contribution of several local industries in Texas who donated most of the materials used in this study.

NOTICE

Contents and Subject Index

I. Introduction

1.1 A Need for This Research

Engineers are currently faced with increasing demands for improved efficiency and reduced concrete construction costs from developers and governmental agencies. As a result, engineers are beginning to design larger structures using higher strength concrete at higher stress levels.

There are distinct advantages in the use of concrete with compressive strengths in the range from 9,000 to 12,000 psi in both reinforced and prestressed concrete construction. For a given cross-section, prestressed concrete bridge girders can carry greater service loads across longer spans if made using high strength concrete. In high-rise buildings, where the main disadvantages of using concrete compared to steel are higher dead loads and large column cross-sections, using high strength concrete makes possible significant reductions in total structural dead weight and in column dimensions. Thus, concrete becomes technically and economically feasible as a structural alternative to steel in tall buildings when high strength concrete is used.

In addition, cost comparisons have shown that the savings obtained through the use of smaller and lighter high strength concrete members are significantly greater than the added cost of the higher quality concrete. Also, observed improvements in durability, shrinkage, and creep characteristics of high strength concrete will decrease serviceability and maintenance problems.

Numerous high strength concrete structures now standing in the U.S. and elsewhere were constructed using concrete with a compressive strength of between 8,000 psi and 11,000 psi. Remarkably, the use of high strength concrete has preceded full information on its engineering properties, which are significantly different in some respects from those of ordinary strength materials. Current understanding of the behavior of concrete under load and the empirical equations now used to predict such basic properties as modulus of elasticity and tensile strength are based mainly on tests of concrete having a compressive strength of about 5,000 psi or less. Extrapolation to higher strength levels is unjustified and may be dangerous. There is an urgent need for studies focussing on the development of constitutive relationships applicable to design of structural members made using high strength concrete. For example, little is known about predicting the material's behavior in high shear zones or its confined strength in overstressed compression members.

Concrete compressive strengths of over 15,000 psi have been achieved in the laboratory for many years. It has been demonstrated that the production of high strength concrete having a compressive strength of 9,000 to 12,000 psi, using conventional materials and production methods, is technically and economicallly feasible [14]. However, very little information has been developed concerning the identification of the most relevant parameters in the selection of materials and their proportions for producing high strength concrete. This is not surprising, given the variability in physical properties and availability of concrete-making materials in different regions of the U.S. Mix design guidelines for high strength concrete need to be developed for each region of the country. Also, current quality control standards, as they relate to materials used in concrete, especially cement, are not narrow enough to ensure consistent production of good quality high strength concrete.

What is needed most is a systematic, reproducible procedure for attaining high strength concrete with readily available materials using conventional ready-mix batching procedures. If an engineer is to take advantage of this material, he must be given reason to be confident that high strength concrete can be produced and used safely, economically, and efficiently. This research program constitutes the much needed first step towards the development of the necessary information for using high strength concrete in highway structures in the State of Texas.

1.2 Objectives

The overall objectives of this research are as follows:

(1) to identify the most relevant properties of cement, aggregate, and admixtures for producing high strength concrete;

(2) to evaluate the suitability of commercially available cements, aggregates, and admixtures in Texas for the production of high strength concrete;

(3) to establish, in a form useful for practicing engineers in Texas, guidelines for the selection of materials and their proportions for producing high strength concrete;

(4) to study the effect of different curing conditions, temperature and relative humidity, typical of those existing in Texas upon the compressive strength of high strength concrete;

(5) to study the effect of mixing temperature and different mixing times typical of those in construction in Texas on the properties of fresh high strength concrete; and

(6) to study the applicability of current methods of measuring concrete strength such as standard concrete cylinder and flexural strength tests in predicting the strength of high strength concrete.

1.3 Definition of High Strength Concrete

High strength concrete refers to concrete which has a uniaxial compressive strength greater than that which is ordinarily obtained in a region. This definition has been widely accepted by practicing engineers because the maximum strength concrete which is currently being produced varies considerably from region to region in the United States.

Further complications in defining high strength concrete arise from specimen types used for compression testing and age at testing. For example, a 6-in. dia. x 12-in. cylinder, as is used in the U.S., and a 4-in. x 4-in. cube, as is used in Europe, molded from the same batch of concrete will yield two completely different compressive strengths. Whether specimens are tested at 28, 56, or 90 days, any of which may be more appropriate than the others for a particular job, can make a tremendous difference in the measured compressive strength.

Researchers and practicing engineers have not yet agreed on what compressive strength constitutes high strength for plain concrete. High strength, normal weight concrete has been defined by some as concrete having a compressive strength of at least 6,000 psi [1, 30, 85] at 28 days. Shah [85] defined high strength for lightweight concrete as having a compressive strength of over 4,000 psi, whereas Albinger [1] set the lower limit for lightweight concrete at 5,000 psi. Others [20, 92] used 8,000 psi as the minimum compressive strength for normal weight high strength concrete. Engineers in the Chicago area who have for some time been using 10,000 psi concrete in high-rise buildings have been developing the technology needed to consistently produce concrete having strengths in excess of 12,000 psi. Perenchio [72] suggested that the upper limit to high strength concrete will not be reached until the strength of the cement paste is fully utilized—at about 25,000 psi.

According to Saucier [82], the eventual ceiling on concrete strength is virtually unlimited. He reported, however, that very high compressive strengths will only be achieved by changing production methods. Currently, he stated, 5,000 to 10,000 psi concrete can be produced nearly anywhere in the U.S. by using conventional production techniques, by properly selecting materials and by maintaining good quality control. It is possible to produce concrete with a compressive strength of up to 15,000 psi by utilizing more expensive materials and improved production techniques. For concrete compressive strengths over 15,000 psi, "exotic" procedures and materials may have to be employed.

The main objective of this research program was to establish criteria for selection of materials and their proportions to achieve uniform, economical, high-quality concrete with a compressive strength between 9,000 and 12,000 psi at 56 days using 6-in. dia. x 12-in. cylinders cast in steel molds. Only ordinary concrete-making materials and conventional production techniques currently used by prestressing plants in Texas were used in this project.

1.4 Applications of High Strength Concrete

There are definite advantages, both technical and economical, in using high strength concrete in structures today. Carpenter [12] listed the advantages of using high strength concrete in highway bridge applications as: (1) greater compressive strength per unit cost, per unit weight, and per unit volume; (2) increased modulus of elasticity which aids when deflection and stability control the design; and (3) increased tensile strength, which is a controlling parameter in the design of prestressed concrete members under service loads. Nilson [84] and Anderson [3] concluded that losses in prestressing forces will be reduced because of improved long-term deflection properties of high strength concrete. The National Crushed Stone Association [35] reported that high strength concrete has greater durability and resistance to abrasion and wear than normal strength concrete. Cracking and damage of precast concrete products during delivery and handling can be reduced by using high strength concrete [29]. Due to a higher fines content, high strength concrete can give a more satisfactory appearance on formed and finished surfaces than normal strength concrete.

It has been estimated that for certain minimum heights and spans of structures, high strength concrete generally permits more economical construction due to reduced structural member cross-section dimensions. This results in a reduction in the volume of concrete required and smaller dead loads.

1.4.1 High-Rise Buildings. Most applications of high strength concrete to date have been in high-rise buildings. High strength concrete has already been used in columns, shear walls, and foundations of high-rise buildings in cities such as Houston, Dallas, Chicago, New York, and abroad. Tall structures whose construction using normal strength concrete would not have been feasible have been successfully completed using high strength concrete. Column and beam dimensions can be reduced resulting in decreased dead weight of the structure, and an increase in the amount of rentable floor space in the lower stories. Reduced dead weight can substantially lessen the design requirements for the building's foundation.

It has been shown [92] that in a 50-story structure requiring 4 ft dia. columns using 4,000 psi concrete, redesign using 8,000 psi concrete would result in a reduction of 33 percent in column diameters. Typically, high strength concrete is used only in columns in the lower stories. It has been suggested that 30 stories is the minimum height for a building for which high strength concrete is beneficial [92].

Nilson [84] stated that despite differences in shrinkage and creep behavior of higher strength concrete used in columns and normal strength concrete used in adjoining slabs, no problems have been encountered in actual structures. Based on material and labor costs and the price of rental space in high-rise buildings in the Chicago area, it was determined that using high strength concrete to obtain the smallest member sections having only 1 percent reinforcement resulted in the most economical construction alternative [1].

The Chicago Task Force [16] reported that 7,500 psi concrete was first used in Chicago in 1965 in the Lake Point Tower. In 1972, concrete having a compressive strength of 9,000 psi was used in the first 20 stories of the 50-story Midcontinental Plaza Building. In 1976, two experimental 11,000 psi concrete columns were instrumented and constructed as part of the River Plaza Project. The tallest concrete structure to date is the 79-story Water Tower Place in Chicago, the first 28 stories of which are supported by 48-in. dia., 9,000 psi tied columns with 8 percent longitudinal reinforcement.

At least two high-rise buildings in New York City have utilized 8,000 psi concrete in the lower story columms: 101 Park Avenue Tower (46 stories) and The Palace Hotel (51 stories). In Toronto's Royal Bank Plaza Project, 8,000 psi concrete was also used.

In Houston, 35 percent of the concrete in the Texas Commerce Tower had a compressive strength of 6,000 psi or greater. Columns, shear walls, and spandrels in the first eight floors were cast using a

7-in. slump, pumped concrete mix which had a 7,500 psi compressive strength [18, 78].

In the 72-story InterFirst Plaza in Dallas, the design strength of the concrete was 10,000 psi [98]. The structure's 16 exterior columns, which vary in size from 6 ft x 6 ft to 8 ft x 8 ft, are set on 30 ft centers and are designed to carry the gravity load and base shear.

1.4.2 Highway Bridges. Prestressed, precast concrete bridge girders in Texas normally do not exceed 135 ft to 150 ft in length. Steel members are currently used for spans greater than 135 ft to 150 ft. High strength concrete would permit using greater spans for a given number of girders, or fewer girders for ordinary spans, than when using normal strength concrete. Carpenter [12] showed that a typical bridge design for a 150 ft span would require using nine girders if 6,000 psi concrete were used while only four girders would be needed if 10,000 psi concrete were used. As a result, the slab thickness had to be increased from 5-1/2 in. to 6-1/2 in. in order to support the traffic load on the wider girder spacing. However, the overall dead load of the bridge was reduced. This comparison was based on allowable tensile stresses in the concrete of $3\sqrt{f'_c}$, an allowable compressive stress of $0.4\ f'_c$, and a live load deflection criteria of L/800, where f'_c refers to concrete compressive strength (psi) and L refers to the girder span. The limiting factor controlling the design in this case was spacing of the prestressing tendons within the girders. The use of fewer tendons of a larger diameter and of new girder sections and shapes may have to be considered for efficient use of high strength concrete in bridge girders.

Japanese I-shaped, box, and rectangular section bridge girders have been constructed using 8,500 psi concrete [62]. These highway and railroad bridges have clear spans of between 100 and 280 ft. The I-girders spanned over 150 ft.

A reduction in number and size of bridge columns and piers can result from a reduction in dead load and use of longer spans due to the use of a higher concrete compressive strength. This will allow for significant savings in cost, labor, and construction time.

Other applications of high strength concrete include both heavily loaded transfer girders and offshore structures [44].

No special or "exotic" techniques were employed in constructing any of the high strength concrete structures mentioned in this section. All utilized high-quality materials and good quality control programs.

1.5 Disadvantages of High Strength Concrete

Most of the disadvantages of using high strength concrete listed by engineers result from a lack of research and available information on the behavior of high strength concrete under actual field conditions. Some of the drawbacks reported in the past have been alleviated by recent developments and improvements in admixtures.

Possible drawbacks in using high strength concrete are listed below [12, 20]:

(1) Increased quality control is needed.

(2) High quality materials are less available and often cost more.

(3) Allowable stresses in codes may discourage the use of high strength concrete.

(4) Minimum thickness or cover may govern the design, preventing realization of full benefit of higher strength.

(5) Total available prestress force may be insufficient to fully develop the strength.

(6) Adequate curing can be difficult due to self-dessication of low water/cement ratio mixes. Even with no water loss by evaporation there is inadequate water for full hydration.

(7) Curing can also be difficult because of the rapidly increasing impermeabiity of high strength concrete, which prevents applied curing water from compensating for any initial moisture loss.

A further disadvantage may be that, in structural members where excessive deflections control the design, full utilization of the material's load-carrying capacity when using high strength concrete would not be possible [12,14]. For instance, the higher flexural strength of a high strength concrete flat slab or plate is of little consequence since deflection often controls design.

1.6 Methods of Producing High Strength Concrete

Several exotic methods for producing high strength concrete have been studied, such as (1) modification with polymers, (2) fiber reinforcement, (3) slurry mixing (preblending water and cement at high speed for efficient hydration), (4) compaction by pressure, (5) compaction by pressure combined with vibration, (6) autoclave curing, and (7) mix proportioning using active or artificial aggregates. One study advocated revibration 2-1/2 hours after initial vibration as a means for achieving higher strengths [49]. Structural design which accounts for additional concrete strength resulting from triaxial compression or concrete confinement is also possible.

However, cost-effective production of high strength concrete in construction today is achieved by carefully selecting, controlling, and combining cement, fly ash, admixtures, aggregates, and water. Freedman [24] stated that in order to achieve higher strength concretes the concrete producer must optimize the cement characteristics, aggregate quality, paste proportioning, aggregate-paste interaction, mixing, consolidation, and curing procedures. The use of fly ash and very low water-cement ratios has been widely recommended for producing high strength concrete.

The National Crushed Stone Association [36] further stated that cooperation and coordination among the engineer, architect, materials suppliers, ready-mix producers, contractor, and the testing and inspection agency are required for a successful high strength concrete project.

1.7 Scope of This Program

This report is divided into six chapters. An introduction and a brief literature review of the production of high strength concrete are presented in Chapters I and II. The experimental work is described in Chapter III. Test results are presented in Chapter IV, and are discussed and analyzed in Chapter V. Conclusions, a cost analysis, and recommendations for producing high strength concrete are presented in Chapter VI.

Approximately 2,500 concrete specimens, representing over 200 different batches of concrete were made and tested as part of this study. While mixing procedures and slump were kept constant, the variables studied include materials, proportions, specimen types, mixing temperature, test age, capping material and curing conditions.

A detailed listing of mix proportion and strength test data for all mixes made is included in Appendix B.

In this study, the research approach was to investigate basic interactions among concrete components in mix proportions which are suitable for producing high strength concrete, i.e., low water-cement ratio and high cement content. For this reason, it was important first to know the effects of using different cements and aggregates in high strength concrete mixes which contained no admixtures, and second, to develop fundamental knowledge regarding other available materials such as fly ash and superplasticizers. Only commercially available materials and conventional production techniques used by the Texas State Department of Highways and Public Transportation were utilized in this program. Valuable guidelines have been established to be followed by practicing engineers in the development of trial mixes for producing high strength concrete. Without question, a trial mix design procedure must be used for proportioning high strength concrete in the field.

II. Literature Review: Mix Proportioning for High Strength Concrete

The following is a survey of technical publications which deal, at least in part, with the production of high strength concrete using conventional production techniques. Materials selection, mix proportioning, and the interaction among these materials are discussed.

2.1 Cement

Proper selection of the cement is one of the most important steps in the production of high strength concrete. For high strength concrete containing no chemical admixtures or fly ash, a high cement content of 8.0 to 10.0 sacks/cu.yd. must be used. For a given set of materials, the optimum cement content beyond which no additional increase in strength is achieved from increasing the cement content must be determined. Albinger and Moreno [1] stated that for any particular combination of materials, an optimum cement content exists above which strength declines and the mix becomes too sticky to handle. Additional cement above the optimum cement content will not compensate for the loss in strength due to the increase in mixing water demand needed in order to make the mix manageable in the field. In Fig. 2.1, the 28-day compressive strength of numerous mixes plotted against cement content is shown [24]. In this case, compressive strength did not increase for cement contents above 8.5 to 10.0 sacks/cu.yd.

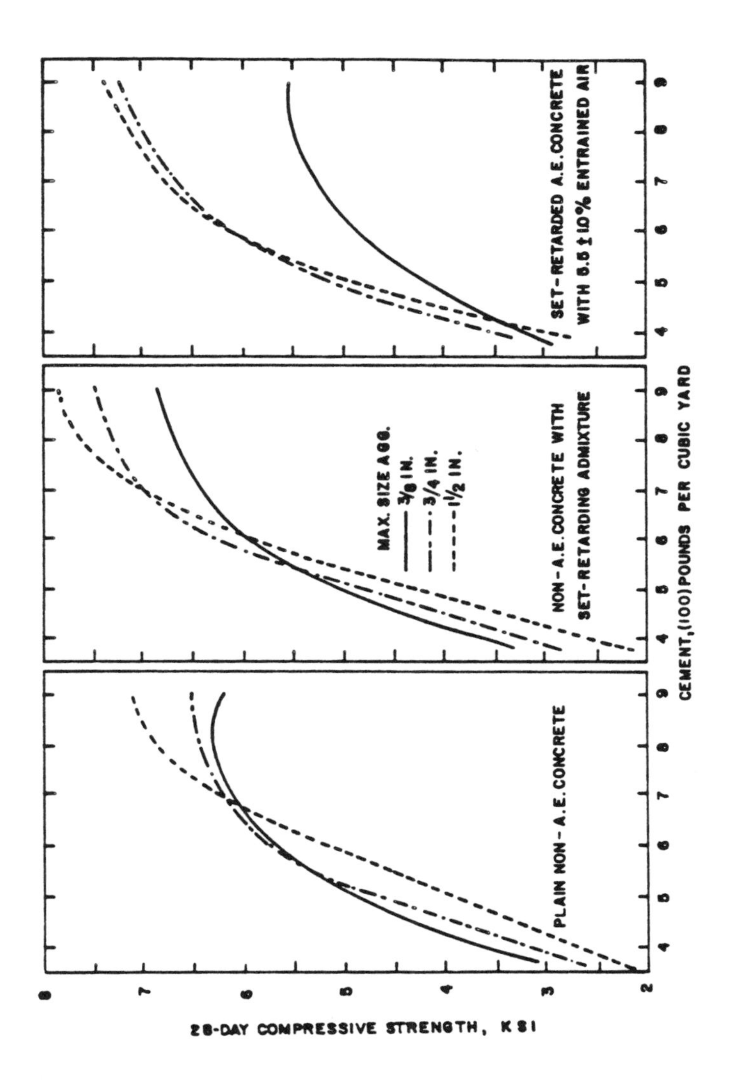

Fig. 2.1 Effect of cement content on the 28-day compressive strength of concrete for various maximum sizes of coarse aggregate in different types of concrete [24].

The Chicago Task Force [16] suggested trial batches using cement contents of 7.0 to 10.0 sacks/cu.yd., comparing strengths on the basis of constant slump. Similarly, Freedman [24] concluded that the cement content must be at least 6.5 sacks/cu.yd. for producing high strength concrete having a 4-in. slump, but that in order to achieve 10,000 psi concrete strengths at 90 days a cement content of 10.0 sacks/cu.yd. is needed. Two studies [20, 105] reported that quantities of cement greater than approximately 9.0 sacks/cu.yd. gave no additional strength. Yamamoto and Kobayashi [105] reported that 9.0 sacks/cu.yd. was the most economical cement content and the minimum for producing high strength concrete without segregation. Another report [88] concluded that the optimum cement content depends on cement type: 10 sacks/cu.yd. for type I cement and 9.25 sacks/cu.yd. for type II cement.

Selection of both type and brand of cement have been shown to be extremely important [16]. Variations in the chemical composition and physical properties of the cement affect the concrete compressive strength more than variations in any other single material. It has been recommended that careful studies be made of variations within one brand and between brands for any area of the country which has plans to produce high strength concrete [16]. These studies should include evaluations of mortar cube strengths in conjunction with concrete trial mixtures. Other studies [1,7] have concurred, and cautioned that the final selection of cement must not be based solely on mortar cube results.

As a result of studies made in Chicago [1, 16], it was recommended that the cement used should provide a minimum 7-day mortar cube strength of at least 4,200 psi. Cement fineness of 4,000 cm^2/g (Blaine) was suggested as a maximum. Another report recommended limiting cement fineness to a maximum of 3,500 cm^2/g to 4,000 cm^2/g (Blaine) for producing high strength concrete [36]. Perenchio [72] found that a much higher early strength was achieved for a cement with a fineness of 10,000 cm^2/g (Blaine), but determined that there was no difference in 90-day strengths between mortars made with the 10,000 cm^2/g cement (Blaine) and one made with a 4,000 cm^2/g (Blaine) cement.

The effects of cement type on strength have been studied also. One study stated that type III cement produced the highest strength concrete for high cement contents up to 90 days after casting. Beyond 90 days, type I cement gave equivalent results [10]. The Chicago Task Force [16] reported that the higher early strengths obtained by using type III cement were not significant in the production of high strength concrete. In mixes made with high cement factors, use of cement types I and II gave early and later age strengths comparable to those of type III cement. This may have been because the type III cement required so much more mixing water for producing concrete with the same slump [16]. Figure 2.2 shows how concrete compressive strength was affected by cement type at high cement contents [88]. It can be seen that using cement types I and II produced higher strength concretes than type III cements, especially for longer curing periods and later testing ages.

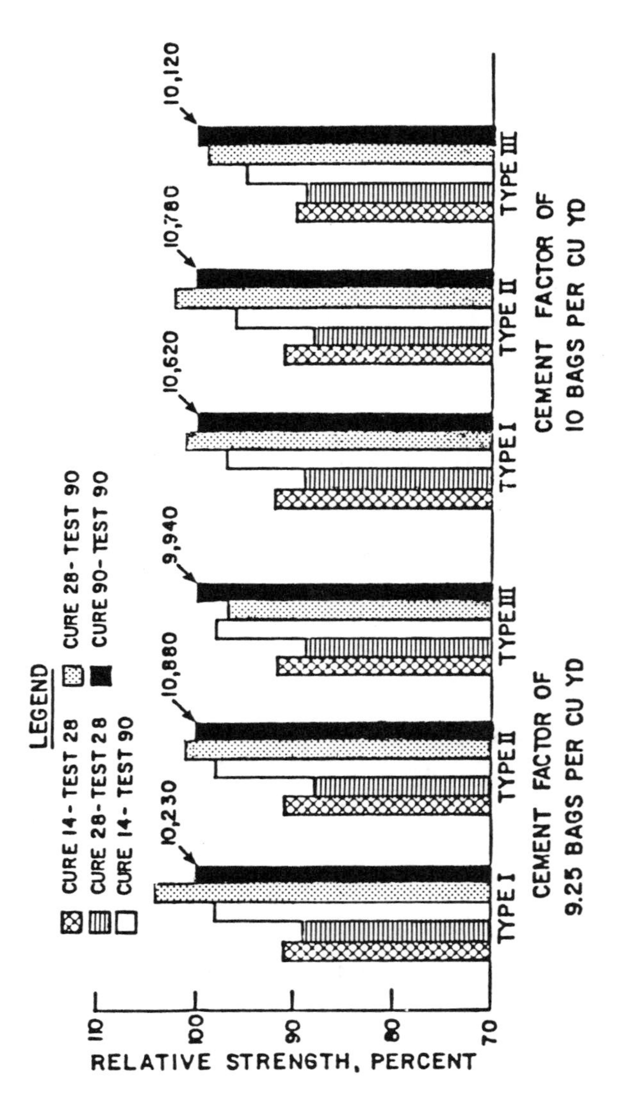

Fig. 2.2 Effect of cement type and curing procedure on the compressive strength of concrete with cement factors of 9.25 and 10 sacks/cu.yd. [88].

For continuous moist curing for 90 days, using type II cement produced the highest strength concrete.

2.2 Water and the Water/Cement Ratio

A U.S. Air Force investigation [88] concluded that the single most important variable in achieving high strength concrete is the water/cement ratio. Others reported [12,72] that the highest concrete strengths were achieved with the lowest water/cement ratios, although considerable effort was required to compact the concrete in some cases. For example, Perenchio [72] acknowledged that the very dry concretes he studied which produced the highest strengths would probably be unacceptable for use in the field in cast-in-place structures.

Most sources agree that high strength concrete cannot be obtained with a water/cement ratio in excess of 0.40. It has been reported that a water/cement ratio in the field of about 0.27 is adequate for hydration of cement [93]. However, others have stated that complete hydration cannot occur with a water/cement ratio of less than 0.38 to 0.40 [90,105]. Concretes having a compressive strength of 9,000 psi to 10,000 psi or more have been produced with water/cement ratios of less than 0.35 in most cases. Figure 2.3 shows the effect of the water/cement ratio on concrete mixes with a constant cement content [24]. In that study, a 90-day compressive strength of 11,000 psi was achieved with a concrete mix which had a water/cement ratio of 0.30 and a slump of 1/2 in.

The difficulty with requiring low water/cement ratios for the production of high strength concrete is overwhelmingly said to be

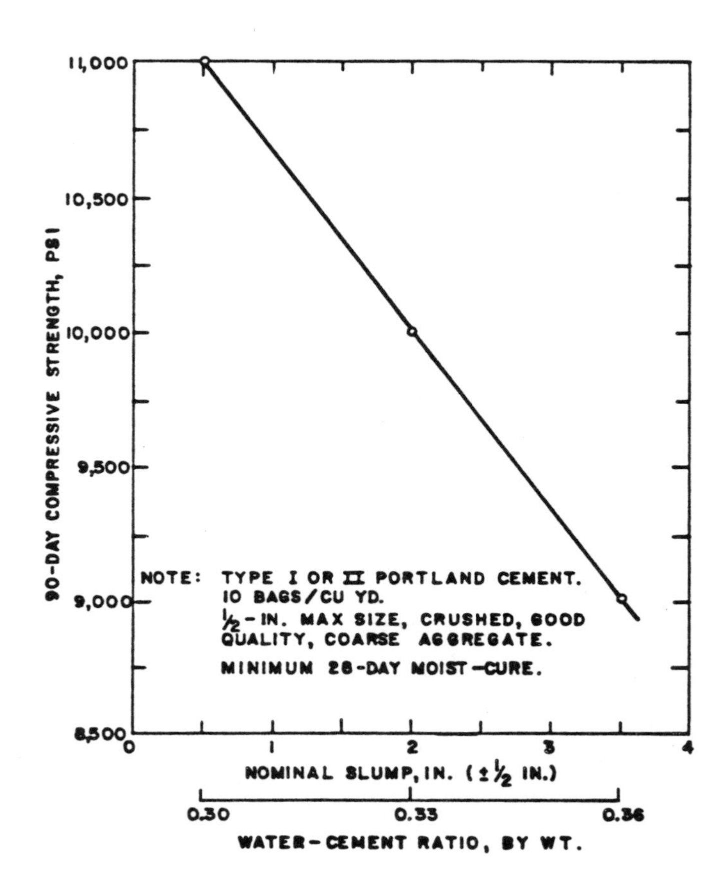

Fig. 2.3 Effect of water-cement ratio and slump on the 90-day compressive strength of concrete [24].

control of water content in the field. Ryan [78] urged close monitoring of moisture content of aggregates and careful control of slump in the field. It was strongly recommended that concrete be delivered on the job with the proper slump so that additional water was not required [1]. When enough water was added to raise the slump by 1 in., Cook [18] reported that at least 250 psi in compressive strength was immediately lost; another study determined that strength was decreased by 700 psi [36] for the same addition of water at the jobsite.

Quality of the water used in concrete is thought to be of no major concern if drinking water is used. Although water temperature affects workability [24,88], it alone will not affect strength significantly. Freedman [24] concluded that unless ice is necessary for hot-weather concreting, the small, if any, increase in strength resulting from the use of ice does not outweigh the problems encountered.

2.3 Coarse Aggregate

Wittman [84] stated that the strengths of aggregates are decisive for determining the ultimate load-bearing capacity of concrete. In ordinary concrete most aggregates have sufficient strength, but, for high strength concrete, aggregates have to be tested carefully. For concretes with strengths of less than 5,000 psi, the aggregate strength is generally greater than the mortar strength. However, for higher strength concrete, the differences in strength and stiffness between the aggregate and the mortar are important parameters [24,68].

Ideal coarse aggregate properties seem mostly to relate to aggregate-mortar bond characteristics and mixing water requirements.

According to Freedman [24], for a constant cement content and maximum aggregate size, differences in the mixing water requirements for a given slump tend to control the strength. Aggregate shape, surface texture, and deleterious coatings are partly responsible for these variations in mixing water requirements. Use of a strong coarse aggregate with moderate absorption has been recommended [72]. Clean cubical, 100 percent crushed stone with a minimum of flat or elongated particles is desirable as well [18, 24, 36]. Freedman [24] advised using an aggregate with an absorption in the range from 1.5 percent to 2.5 percent. He discouraged the use of lightweight aggregate in high strength concrete. The Chicago Task Force [16] stated that mineralogy of the aggregate is also highly important.

Researchers and engineers have agreed that a smaller maximum size coarse aggregate is desirable for high strength concrete. The optimum size for coarse aggregate in concrete depends on the relative strengths of the mortar, the mortar-aggregate bond, and the aggregate particles. For each concrete strength level there is an optimum size for the coarse aggregate that will yield the greatest compressive strength per pound of cement [1,8,16,36]. Use of a 3/4-in. stone has been recommended for producing 7,500 psi concrete, but, for concrete strengths above 9,000 psi, 3/8-in. or 1/2-in. maximum size coarse aggregate is recommended. Since using 1/2-in. coarse aggregate produces a more workable, less sticky concrete mix than using a 3/8-in. stone, 1/2-in. maximum size coarse aggregate is generally recommended for high strength concrete [10,24,72,88]. Reducing the aggregate size to 1/2 in.

in rich mixes has resulted in increases in concrete strength of 10 to 20 percent, even though the water/cement ratio is also increased for a constant cement factor and slump [8,40]. The smaller aggregate size increases the total surface area, thus reducing disruptive stress concentrations and reducing the average mortar-aggregate bond stress [20,24,40,90]. However, Bloem and Gaynor [8] stated that similar aggregates with the same maximum size, but which are from separate sources, may vary more in concrete strength-development characteristics than different sized aggregates from the same source.

The results from one investigation on the effect of maximum size aggregate on concrete strength efficiency are shown in Fig. 2.4 [24]. For a compressive strength of 4,000 psi, the most efficient coarse aggregate size is 1-1/2 in., but using 3/8-in. aggregate is more efficient in producing 7,000 psi concrete. Figure 2.5 shows a comparison of concrete strengths for different sizes of coarse aggregate and different cement factors [24]. In general, it is agreed that smaller size aggregates and higher cement contents produce the highest strengths in concrete mixes with and without admixtures.

Another aspect of coarse aggregate selection which has received considerable attention is the difference in surface texture and particle shape between gravel, or rounded aggregate, and crushed stone. Among the different crushed aggregates that have been studied--traprock, quartzite, limestone, graywacke, granite, and crushed gravel--traprock tends to produce the highest strength concretes [88]. Limestone, however, is more readily available in Texas and in other areas, and

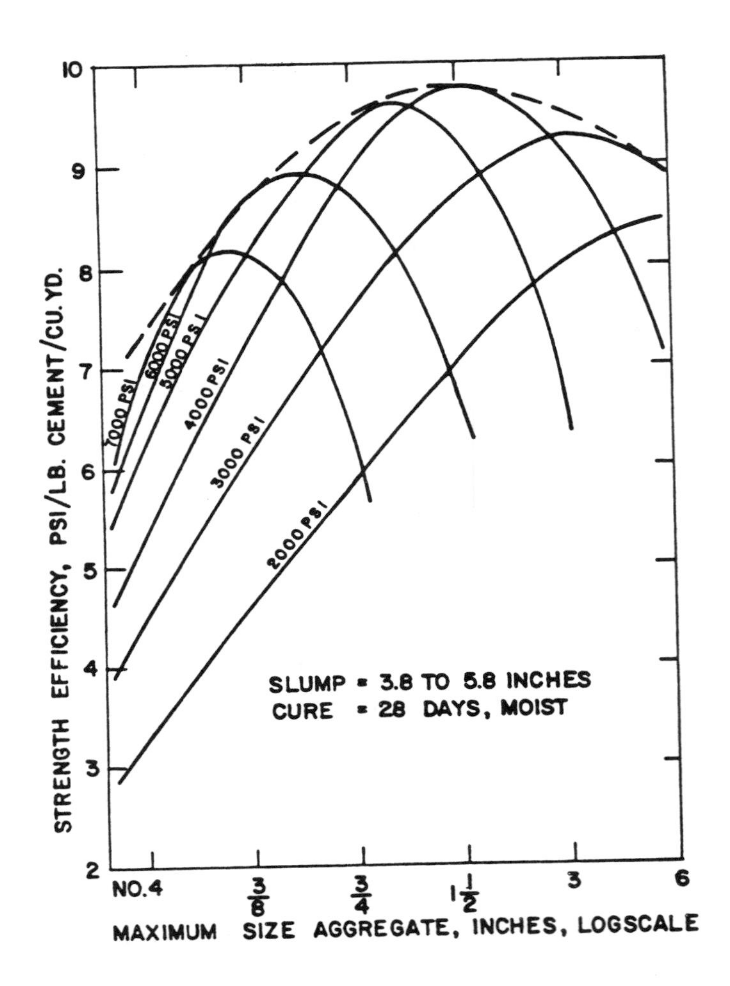

Fig. 2.4 Envelope showing the optimum coarse aggregate size for efficient production of concretes having various compressive strengths [24].

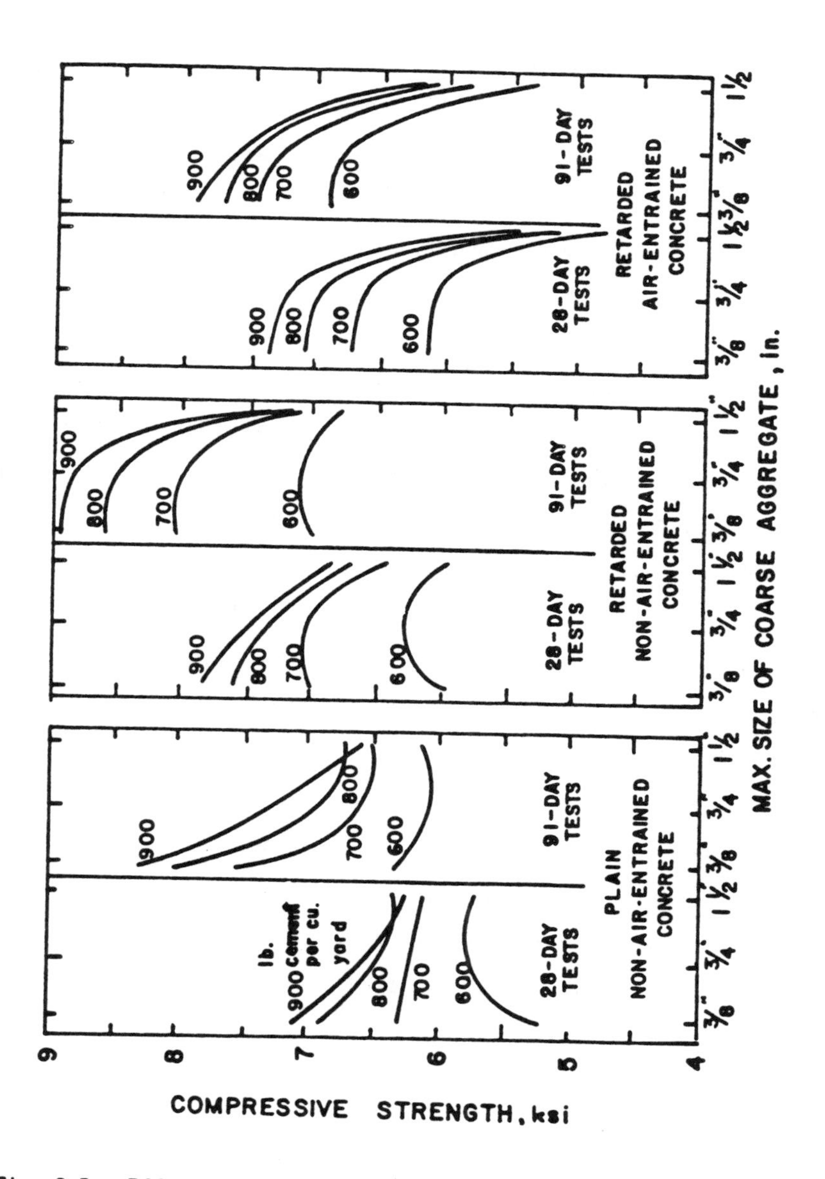

Fig. 2.5 Effect of maximum size of coarse aggregate on the compressive strength of different types of concrete [24].

produces concrete strengths nearly as high as those achieved using traprock. Crushed limestone provides a high aggregate-mortar tensile bond strength in concrete, has a uniform mineralogical composition, and its mineralogical compatibility with the cement matrix may aid in producing high strength concrete [15,20].

However, smooth, rounded coarse aggregates require much less mixing water to obtain a workable concrete. This raises the question of which is more important for concrete strength: the lower water/cement ratio possible when using gravel, or stronger aggregate-mortar bond resulting from the use of crushed limestone. It has been concluded that strength gains from using crushed aggregates overshadow the benefits of increased workability with lower water requirements from using rounded coarse aggregate [24,35]. Carrasquillo et al. [13,15] noted that cracking behavior was similar for gravel and limestone concretes at various strength levels, but that limestone can result in greater ultimate strength, static modulus of elasticity, and ultimate strain. Others [7,16] have also reported a higher strength and static modulus of elasticity for concretes containing crushed limestone.

Gradation of the coarse aggregate within ASTM limits makes very little difference in strength of high strength concrete [16,24,72].

Optimum strength and workability of high strength concrete are attained with a ratio of coarse to fine aggregate above that usually recommended for normal strength concretes [1]. This means using a higher coarse aggregate factor.

The Chicago Task Force [16] recommended using higher coarse aggregate factors than those recommended by ACI Committee 211. Due to the already high fines content of high strength concrete mixes, use of ordinary amounts of coarse aggregate results in a sticky mix.

2.4 Fine Aggregate

Some studies have stated that the fine aggregate gradation is not highly critical for the production of high strength concrete [1,24]. However, it has also been reported that properties of the fine aggregate, especially sand particle shape and texture, have as great an effect on the mixing water requirement of concrete as the properties of coarse aggregate [8]. The fines content in high strength concrete is generally so high due to increased cement contents that using a smaller sand content or a coarser sand is beneficial. Finishability is provided by the high cement content, so that additional fines may only produce stickier, less workable fresh concrete with a greater water demand. Parrott used 10 percent fine aggregate content by weight of total aggregate in producing 11,000 psi concrete [70]. Use of a coarse sand with a fineness modulus in the range between 2.70 and 3.20 has been recommended [1,16,24,83].

One report stated that natural sand is preferable to manufactured, or crushed, sand [88]. The higher mixing water requirement for crushed sand results in lower concrete strengths in spite of the improvement in aggregate bonding characteristics of manufactured sands.

Blending sands for improved capabilities to produce higher strength concrete has also been suggested [90]. If one fine aggregate is detrimental to high strength concrete production, combining it with another different fine aggregate may permit use of the poorer sand in high strength concrete. Blending may aid a ready-mix plant which primarily depends on a source of less desirable fine aggregate for its concrete production.

2.5 Mineral and Chemical Admixtures

The use of mineral and chemical admixtures in producing high strength concrete results in significant increases in concrete strength while reducing the cement requirement and the water/cement ratio. However, the compatability between these admixtures and the cement used must be checked prior to their use in high strength concrete. The fact that a cement, a fly ash, and a chemical admixture individually meet ASTM requirements does not ensure that they are compatible in combination for use in producing high strength concrete [86].

Some concern has been expressed by cement producers that the increasing use of fly ash as a partial replacement for cement in concrete may detract from the demand for cement in this country. On the contrary, the use of fly ash will likely make possible new and unforeseen uses of concrete, resulting in an overall expansion of the market for concrete and cement. This has been the case in the past with the arrival of admixtures such as water reducers [104].

2.5.1 Fly Ash. A good quality fly ash has been said to be mandatory for producing high strength concrete [4, 7, 35, 84]. The concrete strength gain from the use of 10 to 15 percent Class F fly ash, by weight of cement, cannot be attained through the use of additional cement [7, 16]. For Class C fly ash, even higher fly ash contents can be used [1, 18]. However, when using fly ash as cement replacement, by volume or weight, lower compressive and flexural strengths may result at ages less than 90 days [23]. Greater compressive strengths will be achieved at later ages. For comparable early strengths, mixes made with fly ash must contain more fly ash than the amount of Portland cement replaced.

The effect of pozzolans, such as fly ash, on the properties of concrete have been widely investigated, but much controversy still exists about their use in producing concrete [87]. One study demonstrated that 90-day compressive strengths improved when 10 percent of the cement was replaced with fly ash, but concrete strengths dropped when 30 percent of the cement was replaced with fly ash, as shown in Fig. 2.6 [88]. Yamamoto and Kobayashi [105] stated that if any mineral fine, fly ash, blast furnace slag, or even inert standard sand, replaced cement by 15 percent, the strength was essentially unaffected at any age after 7 days, but that replacement by up to 30 percent may cause considerable strength reduction. Another study concluded that replacing 18 to 25 percent of the cement with fly ash, by weight, increases the 28- and 56-day compressive strength and the modulus of elasticity of concrete. Cement replacements with fly ash in the range from 35 to 50 percent resulted in no increase in compressive strength at any age [100]. Two investigations [4,47] reported that fly ash mixes resulted in somewhat lower compressive strengths and elastic moduli at 28 days;

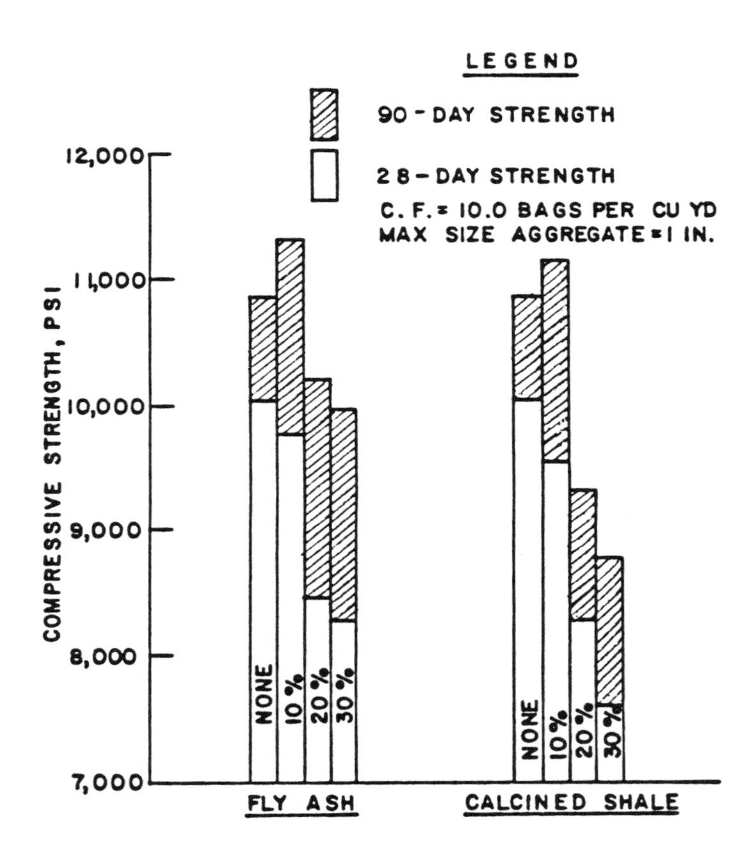

Fig. 2.6 Effect of the amount of pozzolan on the compressive strength of concrete [88].

but the addition of fly ash inevitably resulted in stronger, stiffer concrete at one year of age.

Berry and Malhotra [4] defined a pozzolan as follows: a siliceous or siliceous and aluminous material which itself possesses little or no cementitious value but which will, in finely divided form and in the presence of moisture, chemically react with calcium hydroxide at ordinary temperatures to form compounds possessing cementitious properties. This chemical effect, or pozzolanic action, is secondary in concrete, as it both depends on and follows the hydration of Portland cement [24].

The strength-producing properties of fly ash, a pozzolan, can vary widely. In fact, while the use of some fly ashes results only in pozzolanic action, other ashes contain a cementitious system similar to Portland cement with compounds such as C_3A, C_3S, C_2S, and anhydrite already present in small quantities [57]. The variation in fly ash chemical composition and physical properties is due to differences in the composition of the raw materials used in coal combustion, power plant boiler procedures, and the presence of fuel oil in the combustion chambers in which the ash is produced [57].

Of the 50 to 60 million tons of fly ash produced annually in the U.S., only 15 to 20 percent meets requirements for use in cement or concrete needs [57]. Within that usable portion, there is considerable room for variation in quality and type of fly ash. Therefore, fly ash has been classified into different mineral admixture classes for use in concrete [90]: classes C and F. Class F fly ash is ordinarily produced where anthracite or bituminous coal is burned, which is primarily found in the eastern part of the U.S. Class F fly ash has pozzolanic but no cementitious properties. Class C fly ash is ordinarily produced where lignite or sub-bituminous coal is burned, which occurs primarily in the western part of the U.S. Class C fly ash has both pozzolanic properties and some cementitious properties.

Most fly ashes available in Texas are Class C fly ash. These fly ashes are finer than Class F ashes, are gray to tan in color, and tend to have good strength-gaining characteristics [57,100].

By 1979, fly ash was used in 37 percent of all ready-mixed concrete produced in the U.S. [22]. Over 60 percent of the ready-mix concrete suppliers in the greater Houston area are now reportedly capable of supplying concrete containing fly ash [100].

The best fly ash for use in high strength concrete should have an ignition loss no greater than 3 percent, have a high fineness, and should come from a source whose production quality is fairly uniform [90].

When dealing with high strength concrete, it has been helpful to broaden the "water/cement" ratio concept to include the effect of fly ash on the mixing water requirement [74]. The terms "water/cementitious material" ratio and "water-binder" ratio have been used, where "cementitious material" or "binder" refers to the Portland cement plus all or a portion of the fly ash in the mix.

Benefits from the addition of fly ash to concrete are reported to include increased concrete strength and modulus of elasticity, improved workability and finishability, decreased permeability, reduced heat of hydration, and savings in energy and materials costs [100,104]. Corrosion of reinforcement may be reduced as well [4].

Some possible problems that could arise from using fly ash in high strength concrete include: (1) fly ashes from different origins perform differently in otherwise identical concrete mixes; (2) fly ash may act as a retarder and reduce very early compressive strengths of concrete; and (3) concrete containing fly ash may require more careful curing than plain concrete [74]. Fly ash also reportedly reduces the freeze-thaw resistance of concrete for a given air-entraining admixture dose [47].

2.5.2 High Range Water Reducers (Superplasticizers). Three types of superplasticizers are currently available in the U.S.: (1) a sulfonated melamine formaldehyde condensate which, when added to concrete, forms a lubricating film on the cement particle surfaces; (2) a sulfonated naphthalene formaldehyde condensate, which causes a reduction in the surface tension of the water; and (3) a modified lignosulfate which electrically charges the particles of cement so that they repel each other [40,93]. The net effect of using any type of superplasticizer is enhanced dispersion of cement particles [21]. The initial cement hydration rate is increased since overall water-cement contact is increased. However, the later hydration rate is slower than usual because the reaction product which forms at first around the cement particles tends to be thicker and more impermeable than in non-superplasticized mixes. The film of admixture on hydrating cement particles also tends to restrict further water movement into the cement particles. Some of the admixture apparently even associates with the water on a molecular level, completely preventing a small fraction of the water from ever hydrating the cement [21].

Superplasticizers increase concrete strength by reducing the mixing water requirement for a constant slump, and by dispersing cement particles, with or without a change in mixing water content, permitting more efficient hydration. The addition of superplasticizers to a mix can save cement and increase the slump without changing the consistency of the fresh concrete. High-slump flowing concrete with high compressive strengths have been produced and used which thoroughly fill in the volume surrounding tightly spaced reinforcement, harden quickly to facilitate rapid slip forming, and as a result save 20 to 30 percent in labor cost [31,40,78].

An additional advantage of using superplasticizers results from their use in hot-weather concreting. Slump loss can be successfully readjusted by redosage with superplasticizers instead of with water. A second dosage generally restores the slump and results in greater 28-day strengths [40]. Third and subsequent redoses may not improve strength, but it is important to experiment with higher dosages than those recommended by the admixture manufacturers. Dosage rates as high as 50 percent above manufacturers' recommended amounts have resulted in 10 percent increases in compressive strength without detrimental effects [1,7,16].

The main consideration when using superplasticizers in concrete are the high fines requirements for cohesiveness of the mix and rapid slump loss. Neither is harmful for the production of high strength concrete. High strength concrete mixes generally have more than sufficient fines due to high cement contents. The use of retarders, together with high doses and redoses of superplasticizers at the plant or at the job site can improve strength while restoring slump to its initial amount. Even a superplasticized mix that appears stiff and difficult to consolidate is very responsive to applied vibration [31].

Long-term studies of superplasticized concrete have been conducted in Japan. Test results from 11-year studies showed better strength improvement of superplasticized concretes than of concretes made using a conventional water-reducing admixture or with no admixture at all. Five-year tests showed significantly less corrosion of reinforcement in superplasticized concrete than in control specimens [93].

2.5.3 Air Entrainment. Air entraining agents are not required, nor have they been recommended for high strength concrete in buildings, since the primary applications of high strength concrete, such as caissons, interior columns, and shear walls, will normally not require air-entrained concrete. One investigation recommended that if high strength concrete is to be used under saturated freezing conditions, air entrained concrete should be considered despite the loss of strength due to air entrainment [83]. High strength concrete is much more durable than lower strength concrete; but an air-entrained concrete with only half the strength of high strength concrete is more durable than the high strength concrete containing no entrained air [83]. Ryan [78] stated that effective levels of air content cause an increase in void space which quickly reduces the strength and limits the use of the water/cement ratio as a factor for field control of the mix. It has been shown, however, that adding an air entrained additive to a mix with 2 percent air to get a 5 percent air content reduced the 90-day strength of a 9,400 psi mix by only 2 to 5 percent [83]. In that study, the air entrained mix had a water/cement ratio of 0.03 less than the control mix. This shows that the resulting reduction in the water/cement ratio cannot fully compensate for strength loss due to increased air content. It has been reported that as compressive strengths increase and water/cement ratios decrease, air void parameters improve and entrained air percentages can be set at the lower limits of the acceptable range [90].

2.6 Sample Mix Designs

Listed in Table 2.1 are high strength concrete mix designs from two jobs in Chicago [4,16] and from one study in Texas [18].

2.7 Curing and Testing Considerations

Several variables which have direct impact on the results of concrete compressive strength tests are unrelated to the concrete itself. These other influencing factors are partially responsible for the differences between the strengths of laboratory specimens and field specimens. Variations in results from tests performed on the same material can be caused by differences in specimen shape and size, mold

TABLE 2.1 Sample High Strength Concrete Mix Designs

	River Plaza Building	Water Tower Place	Texas Study [18]
Compressive Strength	11,200 psi @ 56 days	10,600 psi @ 56 days	11,300 psi @ 56 days
Cement (Type I)	850 lb	846 lb	844 lb
Fine Aggregate	1040 lb	1025 lb	765 lb
Coarse Aggregate	1730 lb (1/2-in.)	1800 lb (5/8-in.)	1890 lb (1-in.)
Water	330 lb	300 lb	301 lb
Admixture	43 fl.oz. (retarder)	25.4 fl.oz. (retarder)	32 fl.oz. (reducer)
Fly Ash	100 lb (Class F)	100 lb (Class F)	211 lb (Class C)
Air	1.5%	----	----
Slump	4-1/2 in.	4-1/2 in.	3-3/4 in.
Unit Wt.	148.7 lb/cu.ft.	151.9 lb/cu.ft.	148.2 lb/cu.ft.

materials, methods of consolidation, curing procedures, capping materials and specimen test procedures [30,83].

The age of the specimens when tested is extremely important for high strength concrete. If loading of a high strength concrete bridge girder will not occur until the concrete is at least 90 days old, then the required compressive strength test age could be increased beyond 28 days to take advantage of the 90-day concrete strength in the design of structure [3]. It is very reasonable to specify 90-day strengths in a high-rise building construction since lower floors may not be fully loaded for a year or more [24,78], depending on construction loads. The later age strength criterion may be an additional expense and leave the concrete strength issue in doubt for an uncomfortable length of time in situations of questionable concrete strength [20]. Testing at 90 days of age will typically provide for at least 10 percent greater usable strength compared to 28-day test results [83].

The type of cylinder mold used to cast the compression specimens has a strong effect on compressive strength test results. Rigid steel molds aid in achieving higher and more uniform compressive strength test results due to the more uniform and effective compaction of the concrete and the exactness of standard specimen shape and dimensions which cannot be matched by plastic or cardboard molds [1,30,78]. A steel mold reportedly results in a higher compressive strength test result than does a plastic mold [30]. Using cardboard molds results in compressive strength test results between 2 and 15 percent lower than those of steel-molded concrete [24,30].

Cylindrical specimen size has an effect on concrete strength as well. It was suggested that as specimen size increases, the probability of the presence of a critical flaw in a critical location and orientation likewise increases [83]. Using larger test specimens results in lower average compressive strengths and lower coefficients of variation. Cylinder specimens 6 in. dia. x 12 in. result in an average compressive strength which is 90 percent of that obtained when using 4 in. dia. x 8 in. cylinder specimens [13]. However, one study reported that concrete made with 1 in. coarse aggregate gave the highest strength when using 6 in. dia. x 12 in. specimens, compared to other mold sizes, while concrete made with 3/8 in. stone showed a higher strength when tested using 3 in. dia. x 6 in. cylinder [86].

Curing temperature and humidity affect compressive strength test results in high strength concrete, especially when curing variations occur at early ages. Water curing can add 1,000 psi to the 28-day compressive strength compared to sealed curing. When cured at temperatures above 100°F, variations in water temperatures do not change the concrete strength [83]. Compared with curing at 73°F, curing at 100°F results in higher concrete strengths [83]. Continuous moist curing for 28 days results in 10 percent greater compressive strength and 26 percent greater flexural strength in high strength concrete, compared to specimens moist cured for 7 days followed by curing at 50 to 65 percent relative humidity until testing. Moist curing for 14 days results in about a 5 percent reduction in compressive strength of concrete compared to continuous moist curing [13,70].

Capping thickness and capping compounds have been shown to be important, too. Capping becomes more critical as the strength of the concrete increases [24]. Capping of cylinders must be done with extreme precision using only high strength capping compounds [1,7]. All caps on high strength concrete cylinders must be allowed to develop adequate strength prior to testing [1,7]. Caps with a nonuniform thickness will not transmit the load evenly, and low strength caps may flow or creep under load resulting in induced tensile stresses at the specimen ends [30,90]. Contamination of the capping compound by oil and other impurities must be avoided also [30]. ACI Committee 363 [90] recommended using a 3/8-in. thick high strength cap, having a compressive strength in the range from 7,000 psi to 8,000 psi, or else forming or grinding of all specimen ends. Caps should be allowed to cool for 2 hours, according to Freedman [24].

In addition, testing machines and loading procedures have been shown to cause significant variations in strength. High strength concrete is more sensitive to loading rates than low or moderate strength concretes [30]. When no other information is available, researchers agree that recommended ASTM procedures should be followed when testing high strength concrete.

III. Materials and Test Procedures

3.1 Introduction

High strength concrete is being used increasingly in the field, not only because its production has become economically feasible but also because designers and contractors are slowly beginning to have confidence in its use. Whether or not high strength concrete, especially in the strength range above 10,000 psi, will ever command a significant share of the structural concrete market depends on the ease and consistency with which it can be produced and placed. Although high strength concrete must have a low water/cement ratio, it can be produced using readily available materials and having appropriate workability for ease of placement and proper finishing, even under extreme temperature conditions.

Throughout this investigation, an attempt has been made to include only commercially available materials currently used by precast prestressing plants approved by the Texas State Department of Highways and Public Transportation (TSDHPT). Workability, as measured by the slump test, was the controlling factor for all mixes. All concrete mixes had slumps of at least 3 to 4 inches. Production, curing and testing of concrete specimens in this study were conducted according to applicable procedures described in the TDSHPT Manual of Testing Procedures, Physical Section, 400-A Series, The American Society for

Testing and Materials' *1980 Annual Book of ASTM Standards, Part 14, Concrete and Mineral Aggregates*, and the TDSHPT *1982 Standard Specifications for Construction of Highways, Streets and Bridges*.

In this chapter, a description of the materials, mix proportioning and mixing procedures used in this study are presented.

3.2 Material Properties

The materials used in this study include 5 cements, 5 coarse aggregates, 3 fine aggregates, 2 superplasticizing ASTM type F admixtures, 2 water-reducing and retarding ASTM type D admixtures, 2 sources of fly ash and local tap water. Two or more separate deliveries of several of the materials used were required during the conduct of the study described herein. For this reason, the composition and physical properties of a given material from a single source varied slightly during the course of this study. As a result, each material has a two-part identification number, e.g., A2. The letter represents the source or brand, while the number refers to the delivery date.

With slight, if any, variations in aggregate gradations, as can be seen in the tables shown in Appendix A, the materials used meet applicable TSDHPT and ASTM specifications. Composition and physical properties of the fly ashes and cements used are also presented in Appendix A.

3.2.1 *Cement*. Three cement types, ASTM types I, II, and III, were included in this study. Brands A, B, and D were type I cements. Brands C and E were cement types II and III, respectively. Each of the five cements was produced in Texas at one of four different plants. For mix design purposes, the specific gravity of each cement was assumed to be 3.15.

3.2.2 Coarse Aggregate. Coarse aggregates A through E were all crushed limestones from several aggregate producers in Texas. The maximum size of aggregates B and C was 3/4 in. and 1 in., respectively. The maximum size of aggregates D and E was 1/2 in. Limestone aggregate A was used for a few concrete mixes but became unavailable after the initial delivery. Results from mixes made using coarse aggregate A were limited and incomplete so they are not discussed in this report.

A 1/2-in. maximum size natural gravel, aggregate F, was used for comparison with the limestone aggregates.

Table 3.1 summarizes the properties of these aggregates.

3.2.3 Fine Aggregate. Sands B, C, and D were natural sands (three different sources) having different fineness moduli. The several truck loads of sand B, delivered at three to four month intervals during the study also yielded different fineness modulus values. Fine aggregate A, which was determined to be unacceptable for use in structural concrete was discarded. The few mixes made using sand A are not discussed in this report.

Table 3.2 summarizes the properties of the fine aggregates used in this study.

3.2.4 Chemical Admixtures. Two brands of high-range water reducers, or superplasticizers, ASTM admixture type F, both sulfonated naphthalene formaldehyde condensates, were studied. In calculating the water/cement ratio of mixes containing superplasticizer, the quantity of

TABLE 3.1 Summary of Coarse Aggregate Properties (See Appendix A for more complete descriptions.)

Agg.	Nom. Size (in.)	Texas Grade	ASTM Grade	Type and Description	Bulk Specific Gravity SSD	Absorption (%)	Dry rodded unit weight lb/ft^3
B1	3/4	5	67	crushed limestone (yellow-white)	2.59	2.6	95
B2	3/4	5	67	crushed limestone (yellow-white)	2.63	1.8	96
C1	1	4	57	crushed limestone (yellow-white)	2.57	3.2	99
D1	1/2	7	8	crushed limestone (white)	2.46	4.2	85
E1	1/2	6	7	crushed limestone (gray)	2.65	1.9	97
E2	1/2	6	7	crushed limestone (gray)	2.64	2.1	95
E3	1/2	6	7	crushed limestone (gray)	2.64	1.9	93
E4	1/2	6	7	crushed limestone (gray)	2.68	1.2	95
F1	1/2	7	8	river gravel	2.58	1.5	97
F2	1/2	6	7	river gravel	2.58	0.8	96

TABLE 3.2 Summary of Fine Aggregate Properties (See Appendix A for more complete descriptions.)

Aggre-gate	Fineness Modulus	Bulk Specific Gravity SSD	Absorption	Dry rodded unit weight, lb/ft^3
B1	3.08	2.56	1.0	102
B2	2.57	2.57	1.8	105
B3	2.85	2.57	1.5	107
B4	2.77	2.56	1.7	103
C1	2.72	2.62	1.6	108
C2	2.45	2.64	1.4	104
D1	2.75	2.62	1.0	106

admixture added was included as part of the water. Two water reducer-retarders were (ASTM admixture type D) also used in some mixes.

3.2.5 Fly Ash. Fly ash (A and B) from two different sources in Texas was considered. Fly ash was added to the concrete at a rate of 20 and 30 percent by weight of the Portland cement. Two water/"cement" ratios, by weight, are reported for mixes containing fly ash: "w/c" refers to the ratio of water to Portland cement by weight, and "w/b" refers to the ratio of water to binder by weight. "Binder" refers to the combined weight of Portland cement and fly ash or total weight of cementitious material. Fly ash and Portland cement were batched at the same time.

3.2.6 Water. Tap water was used in all mixes. The unit weight of water was taken to be 62.4 lb/cu.ft. Water temperature was 75°F ± 5° during this study.

3.3 Mixing and Testing

3.3.1 Introduction. All mix designs were based on a saturated surface dry condition of the aggregates. The main variables considered in mix proportioning were: the water/cement ratio required to produce concrete of a given slump, cement factor, and coarse aggregate/fine aggregate weight ratio.

Slump was maintained at 3 to 4 in. in all batches, except those containing superplasticizer. Most mixes containing superplasticizers had slumps in the range from 4 to 5 in. Three cement contents, 7.0, 8.5, and 10.0 sacks/cu.yd. (658, 799, or 940 lb/cu.yd.) were considered.

Coarse aggregate/fine aggregate ratios of 1.0, 1.5, and 2.0 by weight were also considered.

No air entraining admixtures were included in this study.

The concrete was mixed in 3-1/4 cu.ft. batches. For most concrete batches, the following specimens were cast: 6-6 in. dia. x 12 in. cylinders (steel molds), 3 to 6-4 in. dia. x 8 in. cylinders (cardboard and/or steel molds), and 3-6in. x 6 in. x 21 in. flexure text beams. Three 6 in. dia. x 12 in. cylinders from each batch were tested for compressive strength at 56 days. All other specimens, both flexural and compressive strength, were tested at 28 days. Exceptions are noted in the test results in Appendix B. Additional batches were used to study other variables such as: type of cylinder mold including steel, plastic and cardboard molds; effect of high temperature during mixing; mixing time; type of tensile strength specimen, including split cylinder and flexural beam; curing time and conditions. The concrete mixing room including the concrete mixer used in this study are shown in Fig. 3.1.

3.3.2 Mixing Procedures. The mixing procedure for all concrete mixes containing no superplasticizer was to first mix 50 percent of the water with the aggregates followed by the addition of the cement, and then the remainder of the water was added as required to reach the desired slump.

Batches containing superplasticizer were mixed similarly to the mixes without admixture, except that the maximum allowable water/cement ratio was set at 0.30. Slump was then adjusted by adding superplasticizer instead of water. A minimum superplasticizer dose of

Fig. 3.1 Concrete batching laboratory with the concrete mixer used in this study at left.

Fig. 3.2 The 400-kip compressive testing machine used in this study.

6 fl. oz./100 lb cement was added with the initial mixing water to every batch. A limit of 15 fl. oz. of superplasticizer (18 fl. oz. for fly ash mixes) per 100 lb of cement was set to avoid excessive bleeding and extreme retardation effects experienced with lean mixes containing doses of 25 to 40 fl. oz. per 100 lb cement.

A water/cement ratio of 0.30 and an admixture dose of 15 fl. oz./100 lb cement were insufficient to produce the desired slump in some 7-sack mixes, so a water/cement ratio higher than 0.30 was used for those mixes.

For the study of the effect of high temperatures on the properties of fresh concrete, similar batching procedures as described earlier were followed except that the materials were preheated overnight to a temperature of 100°F and hot tap water at a temperature of about 105°F was used for mixing water. During mixing, the mixer was kept hot by continuously running hot tap water over the drum. A plastic cover fitted over the mouth of the mixer prevented cooling of the fresh concrete during the duration of the mixing. Slump was checked at 15 minute intervals. After mixing for 60 minutes, slump was adjusted, if necessary, by adding either water or superplasticizer and three cylinders were cast. After mixing for 90 minutes the slump was again adjusted and three more cylinders were cast. The remainder of the mix was discarded. Mix proportions are given in Appendix B. All other mixes required approximately 15 minutes mixing time before casting.

3.3.3 <u>Tests on Fresh Concrete</u>. The mixer used was a 6 cu.ft. maximum capacity Essex drum mixer with a mixing speed of 30 rev/min.

Concrete was made and molded according to ASTM C192-76, Standard Method of Making and Curing Concrete Test Specimens in the Laboratory, and Tex-418-A, Compressive Strength of Molded Concrete Cylinders, except for the following exceptions from some of the specified procedures:

(1) A primary goal of this research was to show whether or not high strength concrete could be produced with materials presently used by precast prestressing plants. Therefore, coarse and fine aggregates were stored as received, in bins, at a constant moisture content rather than in separate size fractions or under water.

(2) The mixer was moistened thoroughly, but was not buttered before each mix. It is believed that, since this procedure was used constantly throughout this project, it had no effect on relative strength of these mixes.

(3) Except for "hot weather" mixes, every batch was steadily mixed for about 10 to 20 minutes, with stops as necessary to check and adjust the slump until the desired slump was reached.

(4) All 6 in. x 6 in. x 12 in. beams were rodded in three layers, rather than two.

(5) A 5/8 in. dia. rod was used to compact 4 in. dia. x 8 in. cylinders rather than a 3/8 in. dia. rod. This simplified the casting process, since only one rod was needed for all cylinders.

(6) Flexural test specimens were moist cured under the same conditions, 100 percent relative humidity and 73°F ± 3°F, as the compressive strength cylinders.

Slump tests were conducted according to ASTM C143-78, Standard Test Method for Slump of Portland Cement Concrete, and Tex-415-A, Slump of Portland Cement Concrete. The fresh unit weight of every mix was measured according to ASTM C138-77, Standard Test Method for Unit Weight, Yield, and Air Content (Gravimetric) of Concrete, using a 0.10 cu.ft. container. Yield was calculated on the basis of batch weights and specific gravities. As applicable, the Standard Method of Sampling Fresh Concrete, ASTM C172-71, was followed. The temperature of each mix was also recorded.

Specimens were cured in a moisture room meeting ASTM C511-80, Standard Specifications for Moist Cabinets, Moist Rooms, and Water Storage Tanks Used in the Testing of Hydraulic Cements and Concretes.

3.3.4 Testing. With the exceptions mentioned below, the following specifications were followed for compressive, flexural, and split cylinder strength testing: ASTM C39-72, Standard Test Method for Compressive Strength of Cylindrical Concrete Specimens; Tex-418-A, Compressive Strength of Molded Concrete Cylinders; ASTM C78-75, Standard Test Method for Flexural Strength of Concrete (Using Simple Beam with Third-Point Loading); ASTM C496-71, Standard Test Method for Splitting Tensile Strength of Cylindrical Concrete Specimens.

Exceptions to these specifications are as follows:

(1) Nominal specimen dimensions were used in stress calculations and were deemed adequate for the purposes of this project.

(2) The suspended spherically seated block was slightly larger than recommended specifications for the 4 in. dia. x 8 in. cylinders.

Compressive tests were performed using a SATEC 400 kip compression testing machine, shown in Fig. 3.2. Flexure testing was initially carried out on a hydraulic, hand-operated third-point loading beam tester, which has a 12,000 lb capacity. Specimens having a flexural load capacity in excess of 12,000 lb were tested using an Emery testing machine, with a 120 kip capacity. All compressive strength test specimens were capped using Forney's high strength capping compound.

One-day strength tests were conducted between 24 and 27 hours after casting. These cylinders were cured using wet burlap for the first 20 to 24 hours followed by moisture room curing until testing at 24 to 27 hours after casting.

All test results are listed in Appendix B.

IV. Test Results

4.1 Introduction

Experimental test results are presented in this chapter. In Chapter V, the results are discussed and analyzed in relation to the production of high strength concrete.

Chapter IV is divided into sections dealing with the effects of particular component materials or their relative proportions on concrete compressive strength. The effects of compression cylinder mold type and size on the measured compressive strength of high strength concrete are presented. Flexure beam and split cylinder test results are also included. In addition, observations on the workability of fresh concrete mixes containing high dosages of superplasticizers, high coarse aggregate contents, and fly ash are reported.

In this study, the research approach was to investigate basic interactions among concrete components in mix proportions which are suitable for producing high strength concrete. The effects of aggregate type and gradation, and cement type and brand on concrete compressive strength were initially studied in concrete mixes containing no admixtures. Later, superplasticizers and fly ash were added to the mix proportions. The results presented apply to the specific materials used in this study. Changing the materials can be expected to affect the results somewhat.

The term "w/c" refers to the ratio by weight of water to Portland cement; the ratio of coarse aggregate to fine aggregate by weight is referred to as "CA/FA".

All compressive strengths reported are average values of at least three 6 in. dia. x 12 in. cylinders cast using steel molds unless otherwise noted.

4.2 Cement Content

Nearly all mixes studied contained either 7.0, 8.5, or 10.0 sacks (660, 800, or 940 lbs) of cement per cubic yard of concrete. With very few exceptions, 10-sack mixes containing no chemical or mineral admixtures resulted in greater compressive strengths than either 7- or 8.5-sack mixes, regardless of mix proportions. The relationship between 56-day concrete strength and cement content for a concrete mix made using type II cement, 1/2 in. crushed limestone E and sand B, is shown in Fig. 4.1. Compressive strengths of approximately 9,500 psi were obtained at 56 days using 10.0 sacks of cement per cubic yard, while mixes containing 8.5 sacks/cu.yd. reached only about 8,500 psi. Typical compressive strength results are plotted versus specimen age in Fig. 4.2. As shown in Fig. 4.3, for mixes containing no admixtures the higher the cement content, the higher the compressive strength for any type of cement.

In general, the optimum cement content of high strength concrete mixes containing superplasticizers was 8.5 sacks/cu.yd., regardless of mix proportions used in this study and specimen age, as shown by Figs. 4.4 and 4.5.

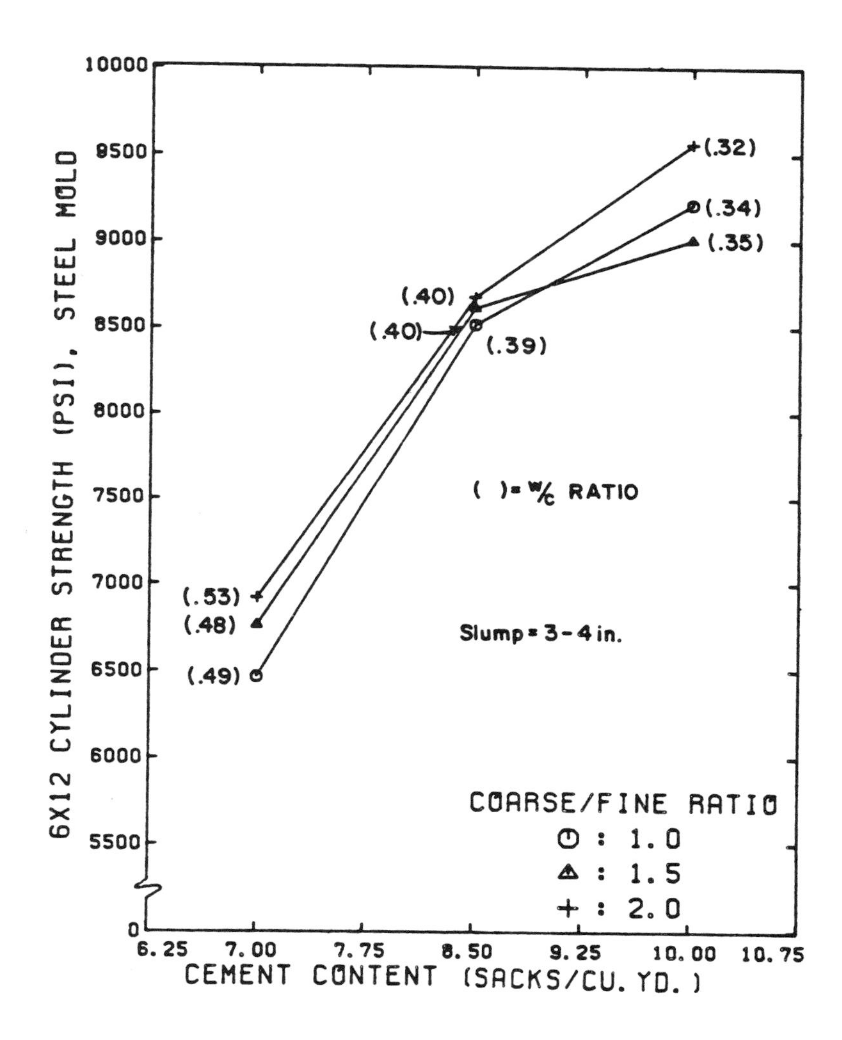

Fig. 4.1 Effect of cement content and CA/FA ratio on the 56-day compressive strength of concrete for mixes made with type II cement, 1/2-in. limestone E, sand B, and no admixture.

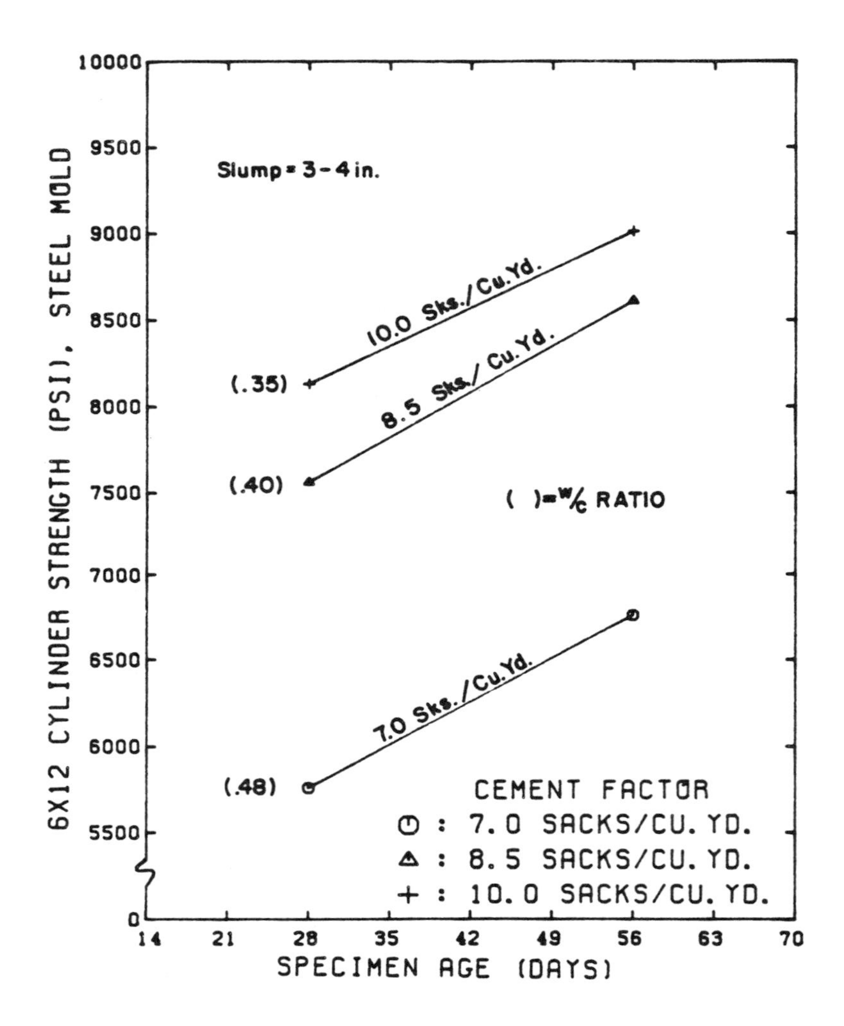

Fig. 4.2 Effect of specimen age and cement content on the compressive strength of concrete for mixes having a CA/FA ratio of 1.5 and made with type II cement, 1/2-in. limestone E, sand B, and no admixture.

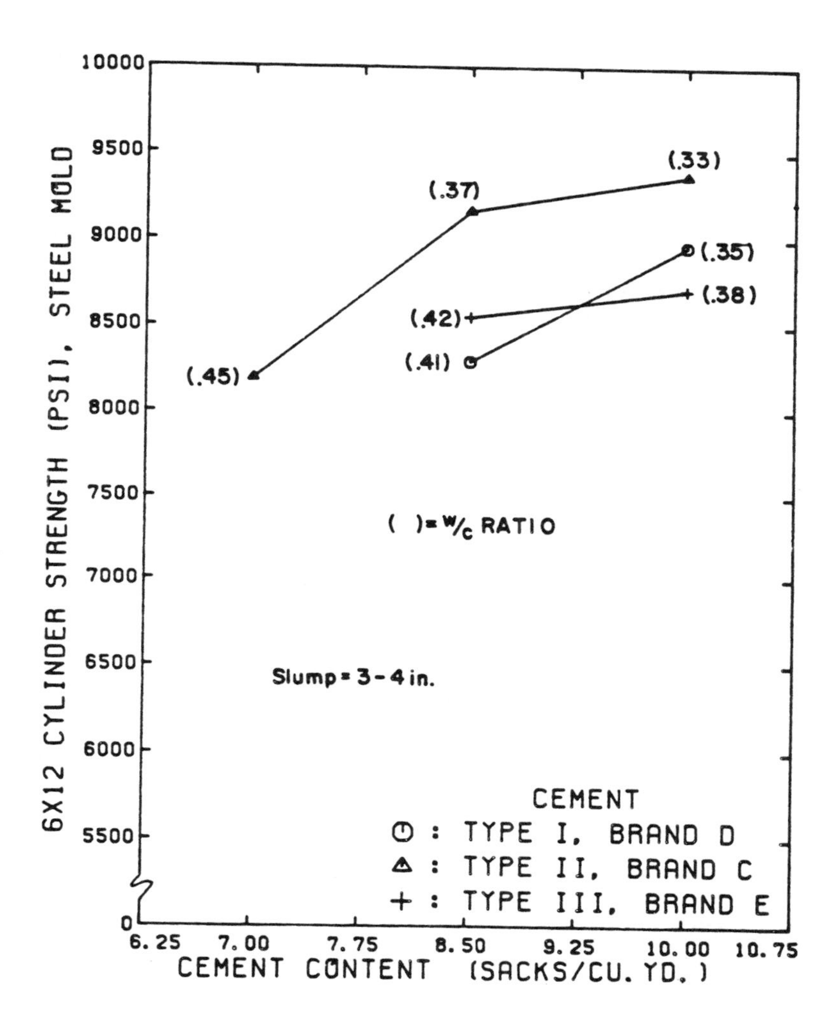

Fig. 4.3 Effect of cement type and cement content on the 56-day compressive strength of concrete for mixes having a CA/FA ratio of 1.5 and made with 1/2-in. limestone E, sand C, and no admixture.

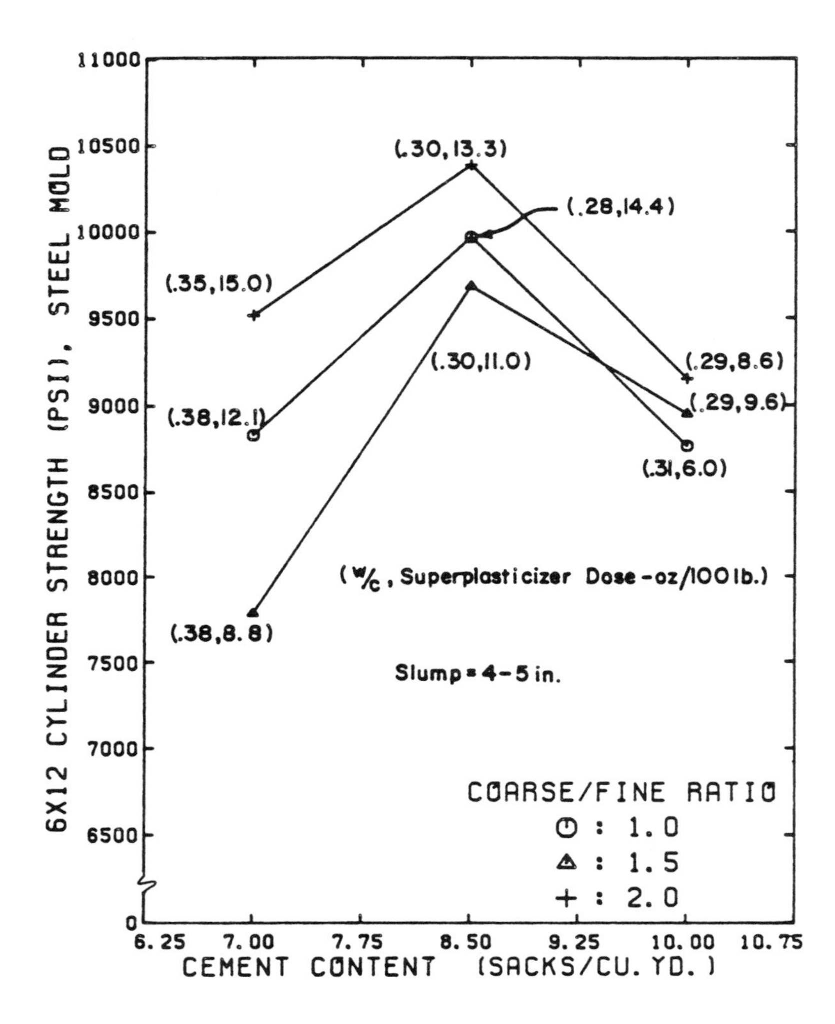

Fig. 4.4 Effect of cement content and CA/FA ratio on the 56-day compressive strength of concrete for mixes made with type II cement, 1/2-in. limestone E, sand C, and superplasticizer B.

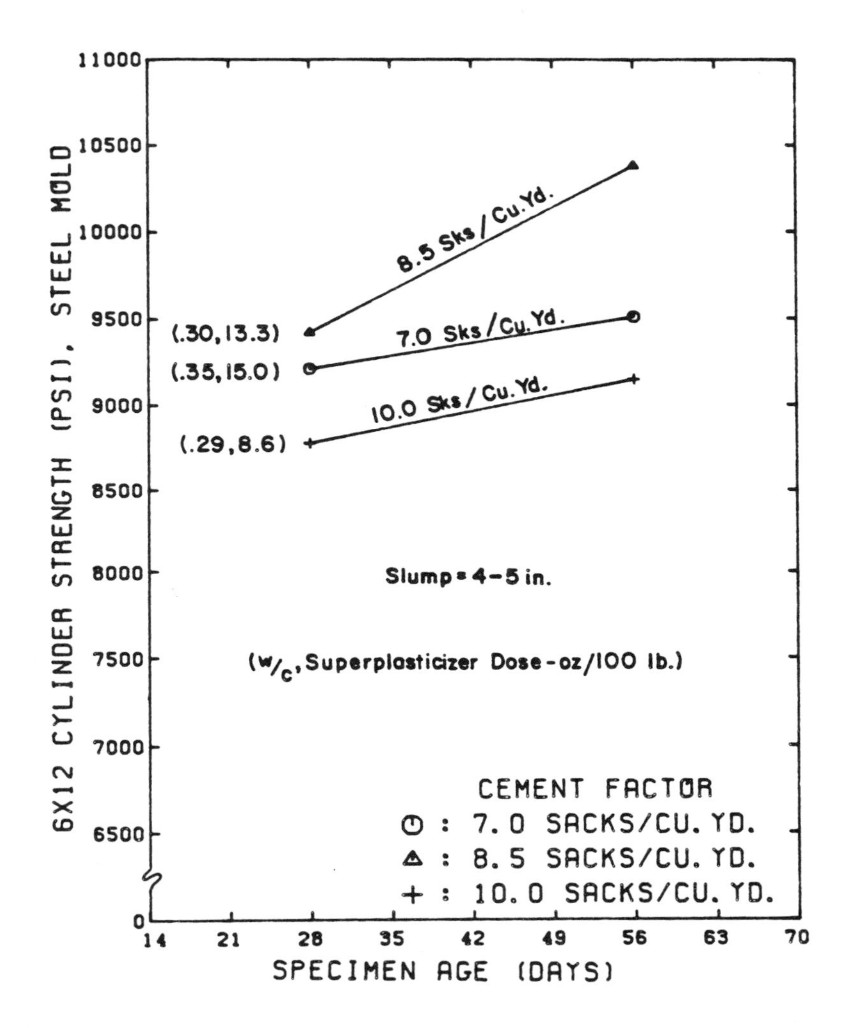

Fig. 4.5 Effect of specimen age and cement content on the compressive strength of concrete for mixes having a CA/FA ratio of 2.0 and made with type II cement, 1/2-in. limestone E, sand C, and superplasticizer B.

4.3 Water/Cement Ratio

The w/c ratio was the most influential parameter affecting the compressive strength of high strength concrete mixes in this study.

As shown in Figs. 4.6 and 4.7, it is clear that the lower the water/cement ratio, the higher the compressive strength, regardless of test age, materials used, and mix proportions. The scatter of data observed in these figures results greatly from not considering the weight of fly ash contained in many high strength concrete mixes as part of the weight of cement. However, if the compressive strength of the concrete is plotted versus the water/binder ratio (w/b) where binder refers to the total weight of cement plus Class C fly ash, a much better correlation is observed, as shown in Figs. 4.8 and 4.9. These figures include results of all compressive strength tests in this study. From these figures it is clear that a low w/b or w/c ratio is of primary importance for producing high strength concrete regardless of specimen age, and materials and mix proportions used.

4.4 Cement Type

Early in the research program, numerous comparisons were made among different brands of type I cement, and among three cement types, I, II, and III, in concrete mixes made using 3/4-in. crushed limestone, sand B, 8.5 sacks of cement/cu.yd. and no admixtures. For these mixes, compressive strengths were less than 9,000 psi. As shown in Fig. 4.10, there was a definite effect on concrete compressive strength of the brand of type I cement used independent of aggregate proportions and water-cement ratio.

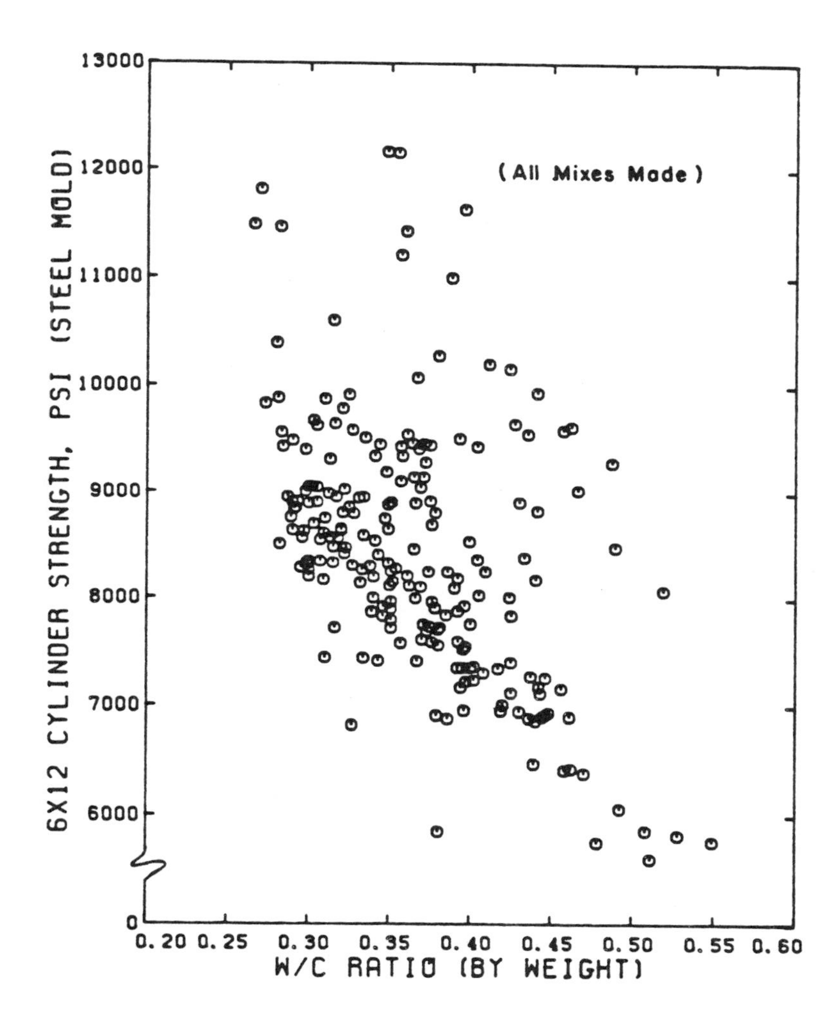

Fig. 4.6 Effect of water-cement ratio on the 28-day compressive strength of concrete for all 6-in. dia. x 12-in. cylinder specimens made, with and without chemical admixtures and fly ash.

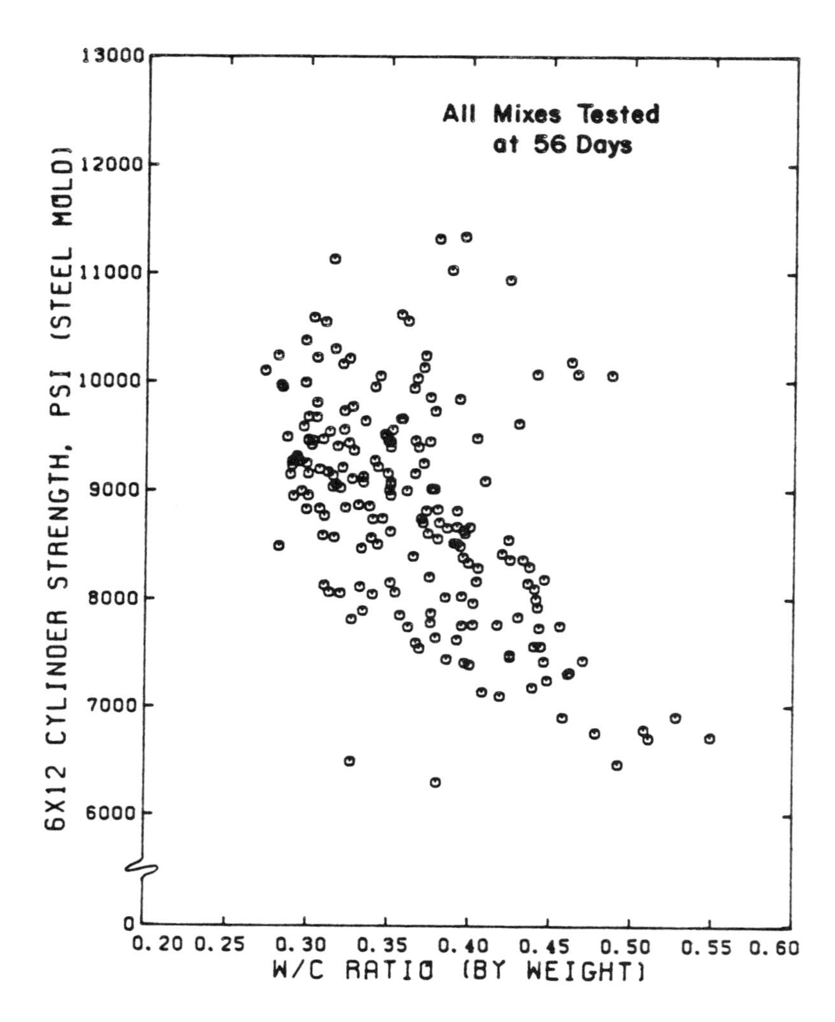

Fig. 4.7 Effect of water-cement ratio on the 56-day compressive strength of concrete for mixes made with and without chemical admixtures and fly ash (6-in. dia. x 12-in. cylinder specimens).

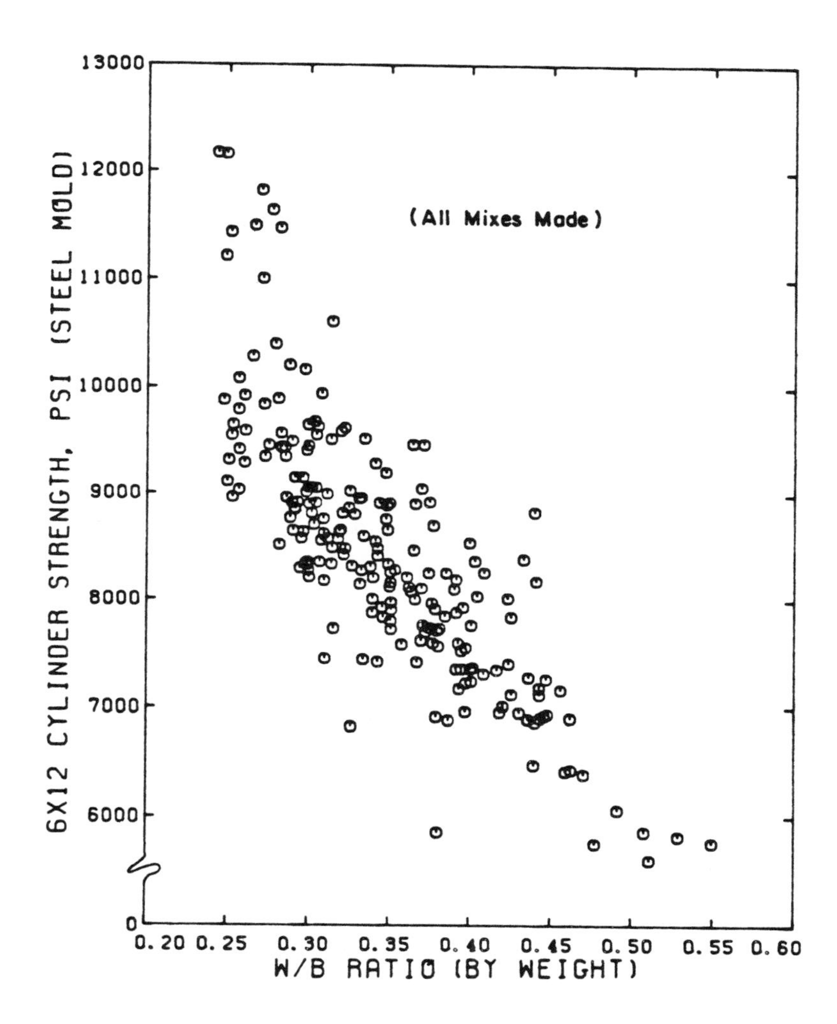

Fig. 4.8 Effect of water-binder ratio on the 28-day compressive strength of concrete for all 6-in. dia. x 12-in. cylinder specimens made, with and without chemical admixtures and fly ash.

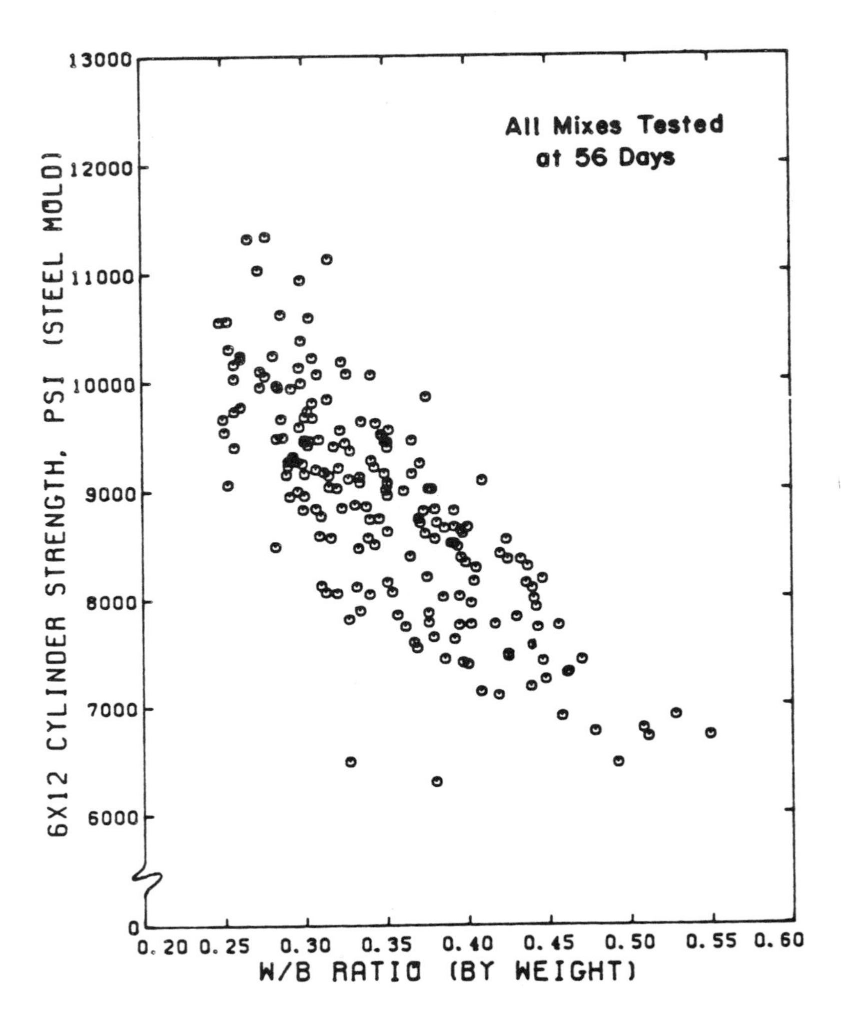

Fig. 4.9 Effect of water-binder ratio on the 56-day compressive strength of concrete for mixes made with and without chemical admixtures and fly ash (6-in. dia. x 12-in. cylinder specimens).

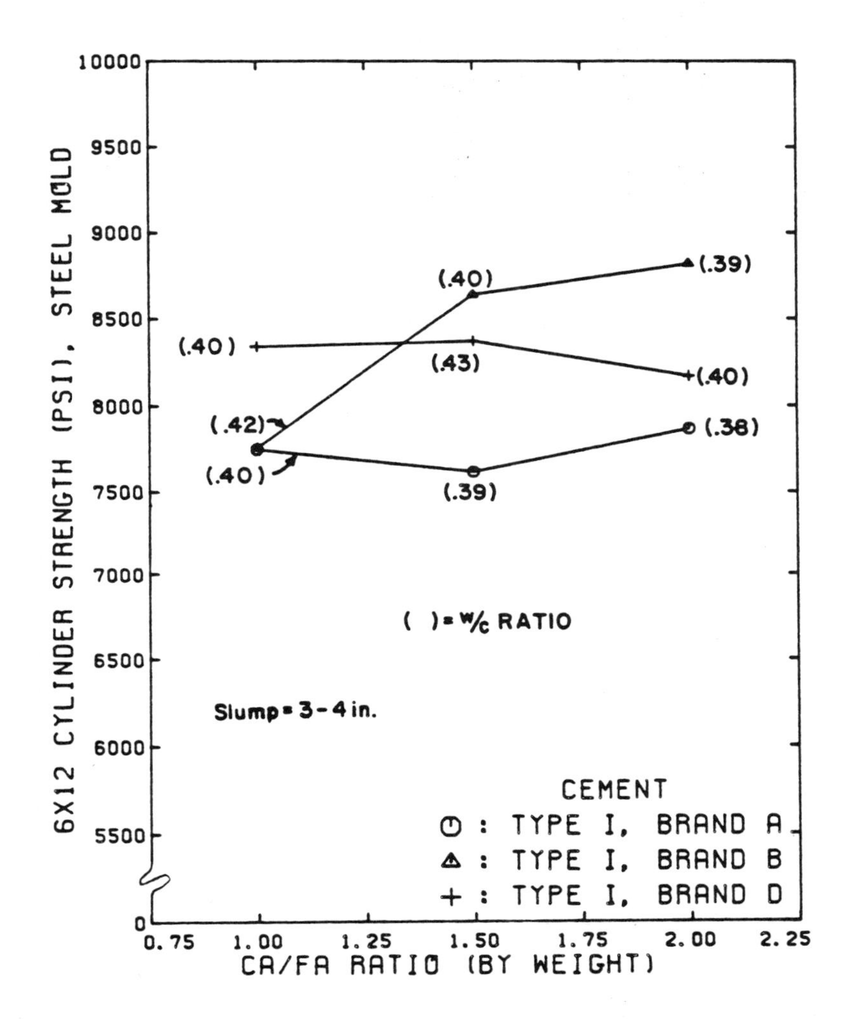

Fig. 4.10 Effect of brand of type I cement and CA/FA ratio on the 56-day compressive strength of concrete for mixes made with 8.5 sacks of cement per cu.yd., 3/4-in. limestone B, sand B, and no admixture.

The effect of cement type on concrete compressive strength was more significant in high strength concrete mixes than in concrete having a compressive strength less than 9,000 psi. In concrete mixes made using 3/4-in. and 1-in. maximum size coarse aggregate which resulted in compressive strengths of less than 9,000 psi, the use of type II cement did not result in higher concrete strength than that obtained when using type I or III cement regardless of testing age and aggregate proportions, as shown in Figs. 4.11 and 4.12. This was true despite the lower water demand of type II cement.

However, for mixes made with 1/2-in. maximum size coarse aggregate, as shown in Figs. 4.13 and 4.14, use of type II cement resulted in higher compressive strengths than did the use of other types of cement, regardless of cement content, aggregate proportions, testing age, and sand fineness. As shown in Fig. 4.14, concrete compressive strengths achieved in mixes containing no admixture using type II cement were greater than 9,000 psi, which is nearly 10 percent greater than the compressive strength of concrete mixes made using types I and III cements. As shown in these figures, less mixing water was required for a given workability in high strength concrete mixes containing type II cement than in mixes made with types I and III cements.

Figure 4.15 shows that for high strength concrete made using a superplasticizer, type II cement produced the highest concrete strengths regardless of aggregate proportions. In addition, type II cement required less mixing water and less admixture than type I and type III cements to achieve a 4 in. slump. The type II cement used also produced

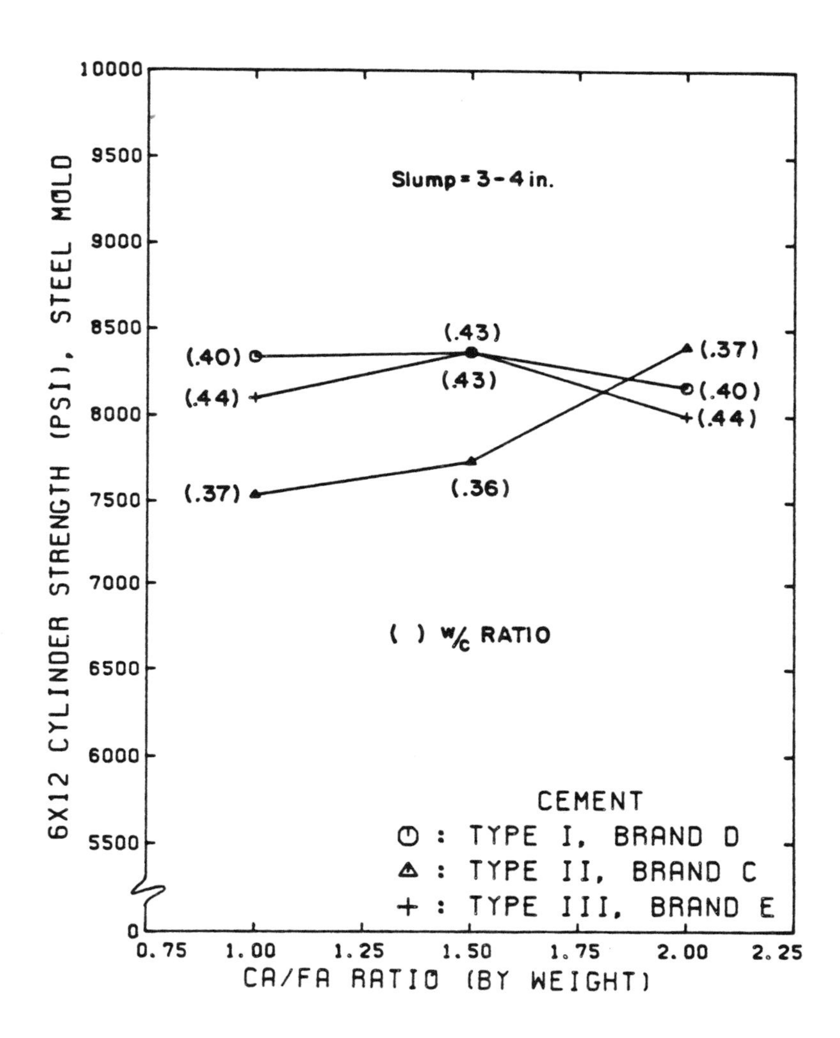

Fig. 4.11 Effect of cement type and CA/FA ratio on the 56-day compressive strength of concrete for mixes made with 8.5 sacks of cement per cu.yd., 3/4-in. limestone B, sand B, and no admixture.

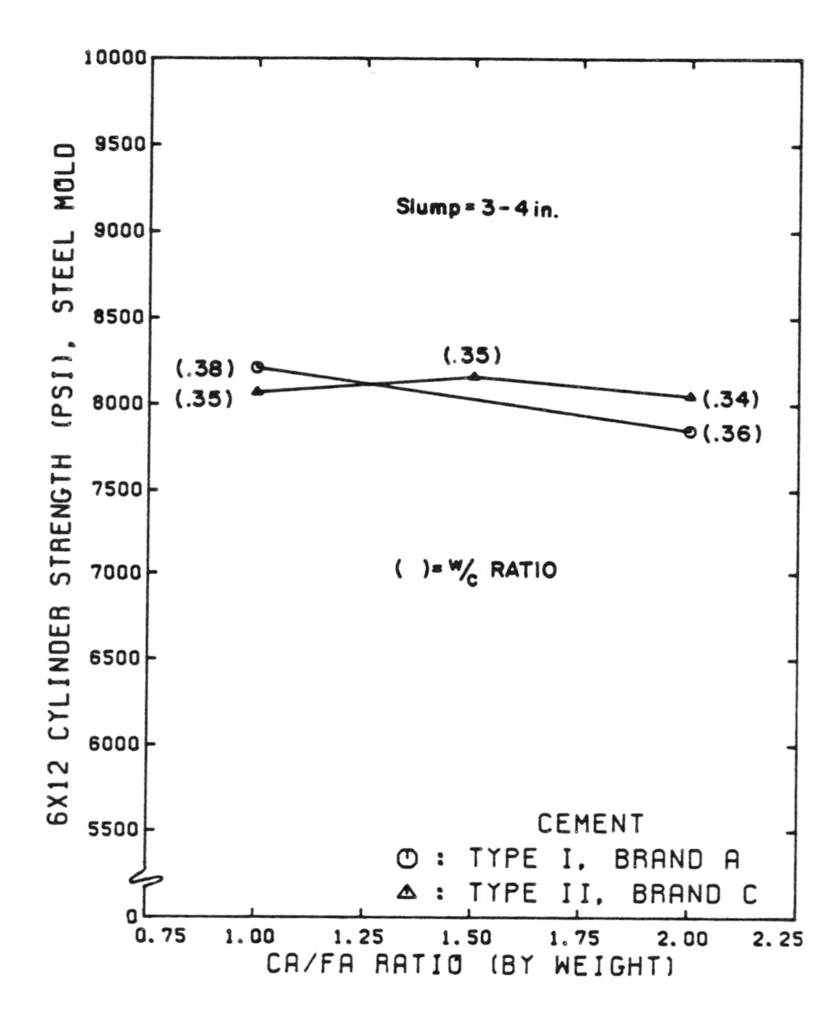

Fig. 4.12 Effect of cement type and CA/FA ratio on the 56-day compressive strength of concrete for mixes made with 8.5 sacks of cement per cu.yd., 1-in. limestone C, sand B, and no admixture.

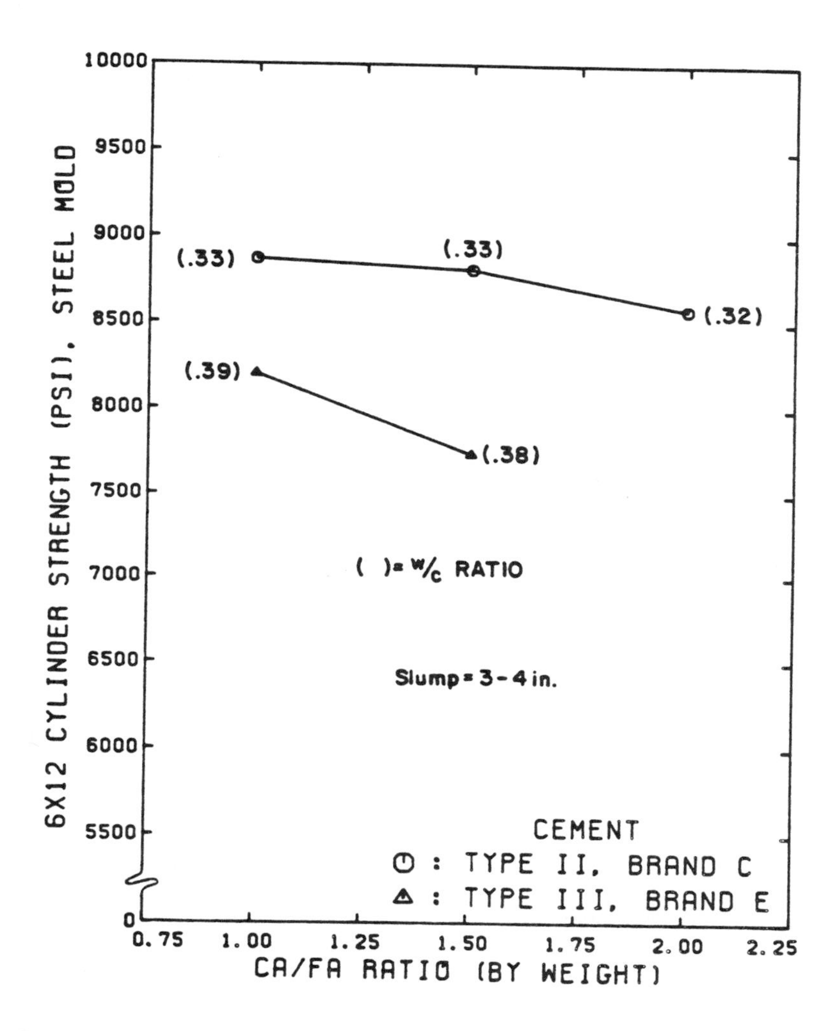

Fig. 4.13 Effect of cement type and CA/FA ratio on the 28-day compressive strength of concrete for mixes made with 10 sacks of cement per cu.yd., 1/2-in. limestone E, sand C, and no admixture.

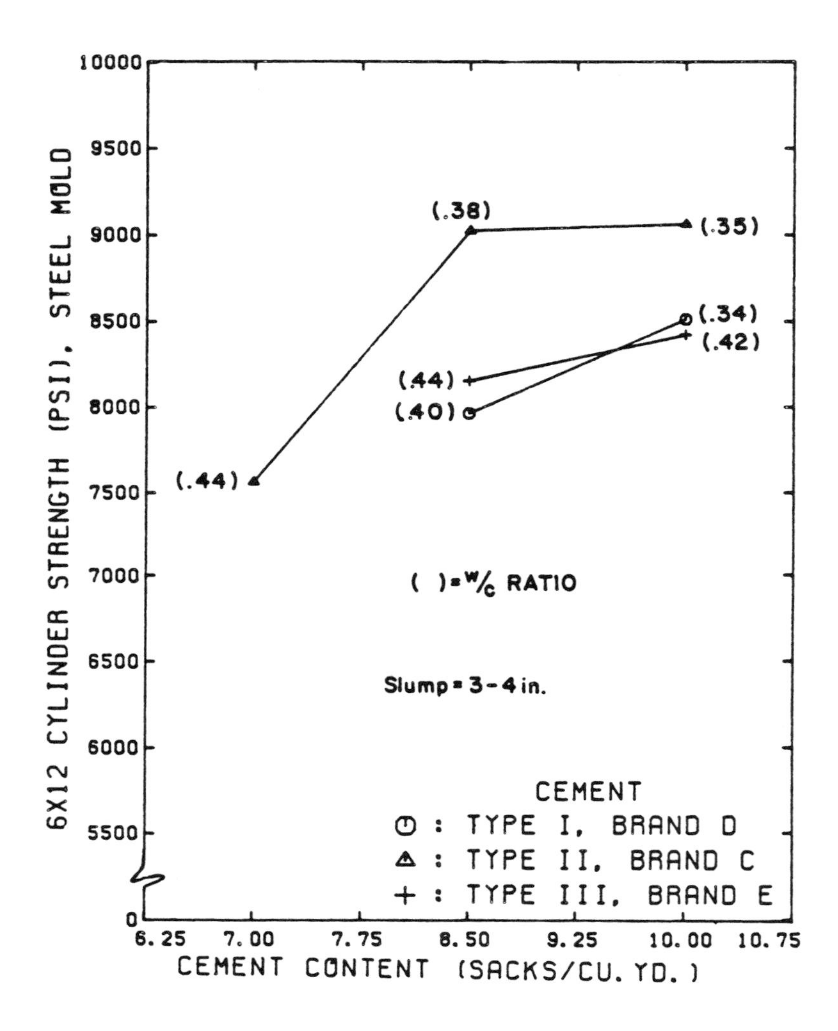

Fig. 4.14 Effect of cement type and cement content on the 56-day compressive strength of concrete for mixes having a CA/FA ratio of 1.5 and made with 1/2-in. limestone E, sand D, and no admixture.

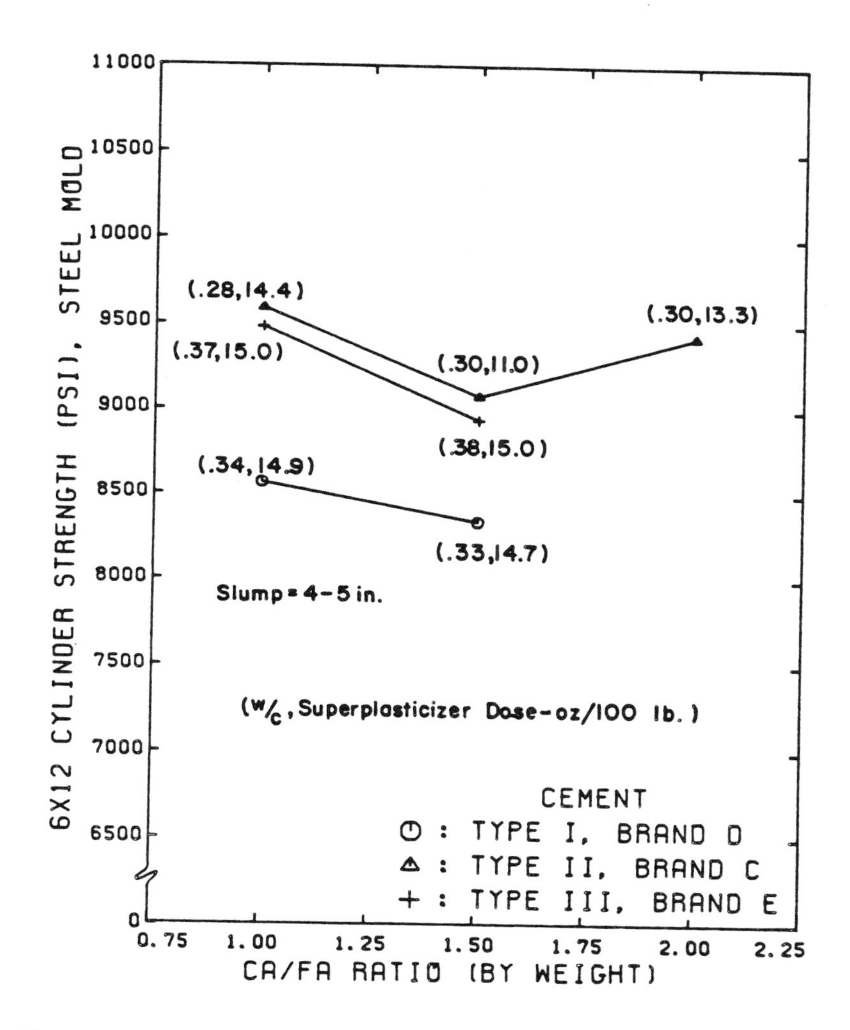

Fig. 4.15 Effect of cement type and CA/FA ratio on the 28-day compressive strength of concrete for mixes made with 8.5 sacks of cement per cu.yd., 1/2-in. limestone E, sand C, and superplasticizer B.

mixes which had better workability than the concretes made using the type I and type III cements.

It can also be seen in Fig. 4.15 that mixes having a CA/FA of 1.5 had slightly lower concrete strengths than mixes made with either greater or smaller CA/FA ratios, regardless of cement type. It was noted frequently in series of mixes containing superplasticizers that the lower concrete strengths tended to correspond to mixes which required a lower dosage of superplasticizer for the same slump. The 8.5-sack mixes having a CA/FA ratio of 1.0 were generally sticky and needed a greater admixture dose to achieve a 4-in. slump, for a given water/cement ratio, compared to the mixes having a higher CA/FA ratio. On the other hand, concrete mixes having a CA/FA ratio of 2.0 were rocky in texture compared to mixes having a lower CA/FA ratio, so a higher superplasticizer dose was required to reach a slump of 4 in. at a given w/c ratio. In general, a CA/FA ratio of 1.5 in 8.5-sack mixes resulted in slightly lower compressive strengths but produced the most workable mix requiring the lowest admixture dose. Further addition of superplasticizer above that needed to produce a 4-in. slump at a w/c ratio of 0.30 was not investigated. Using as much superplasticizer as a mix can hold without workability or segregation problems could result in both higher strengths and higher slumps.

4.5 Superplasticizer Dose and Brand

It was generally observed that for two identical high strength concrete mixes having the same w/c ratio, the one with a higher superplasticizer dosage produced concrete with higher compressive strength.

This was particularly true for mixes which had cement contents of at least 8.5 sacks/cu.yd. It was also noted that a lean, rocky mix was rendered harsh and segregated by the addition of high dosages of admixture. This type of a mix was also exceedingly slow to set, unfinishable, and weaker at any test age.

The relationship between compressive strength and superplasticizer dosage for all mixes tested is shown in Figs. 4.16 through 4.19. Superplasticizer dosage is expressed in fl.oz./100 lb of cement in Figs. 4.16 and 4.17 and as a percent by weight of the total mixing water in Figs. 4.18 and 4.19. The point labelled "A" in each of these figures corresponds to a lean, high-dosage concrete mix which hardened at such a slow rate that it could not be removed from the molds until 48 hours after casting. The typical effects of brand and dosage of superplasticizer on concrete compressive strength are illustrated in Fig. 4.20. As explained in Chapter III, an attempt was made to maintain the w/c ratio at 0.30 and the superplasticizer dosage between 6 and 15 fl.oz. per 100 lbs of cement. Additional water in excess of that corresponding to a w/c ratio of 0.30 was added if the slump was inadequate with an admixture dose of 15.0 fl.oz. per 100 lbs of cement. As shown in Fig. 4.20, additional water above a w/c ratio of 0.30 was generally required for the 7-sack mixes. A lower w/c ratio could have been achieved with higher admixture dosages but it was not tried in this study. As is also seen in Fig. 4.20, significant strength increases of approximately 25 percent occurred in 7-sack mixes when, because of the addition of either brand of superplasticizer, the w/c ratio was reduced from 0.46 to 0.38.

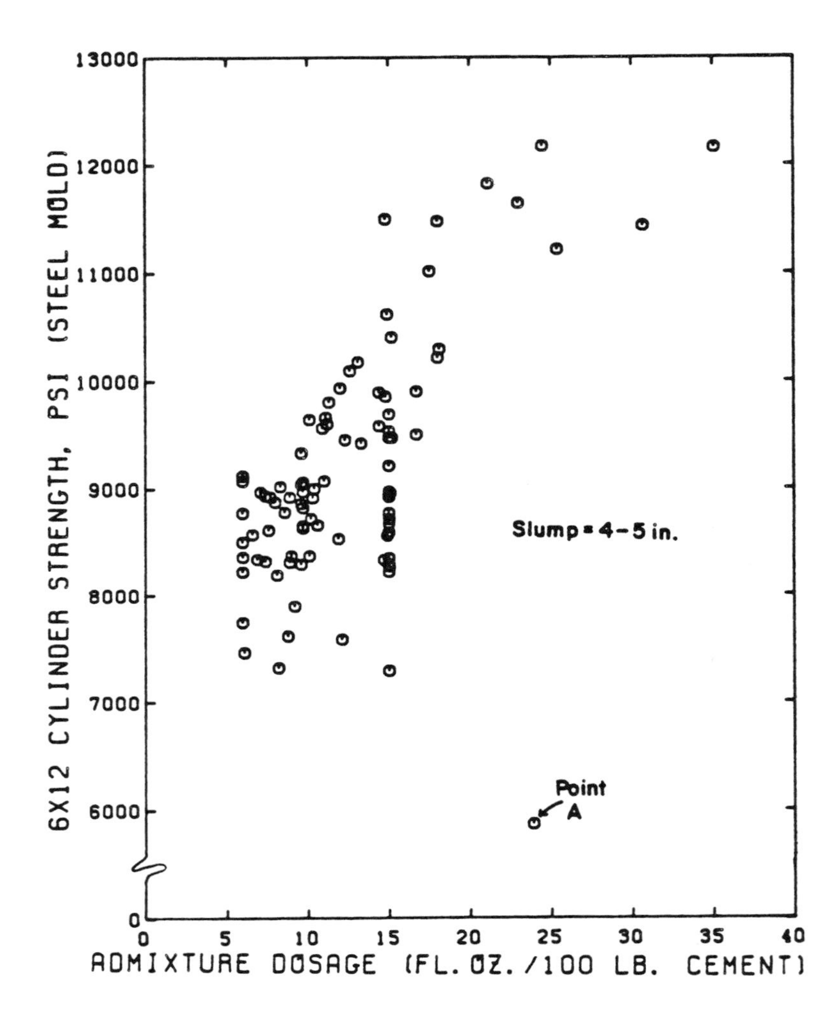

Fig. 4.16 Effect of superplasticizer dosage on the 28-day compressive strength of concrete for all mixes made containing superplasticizer.

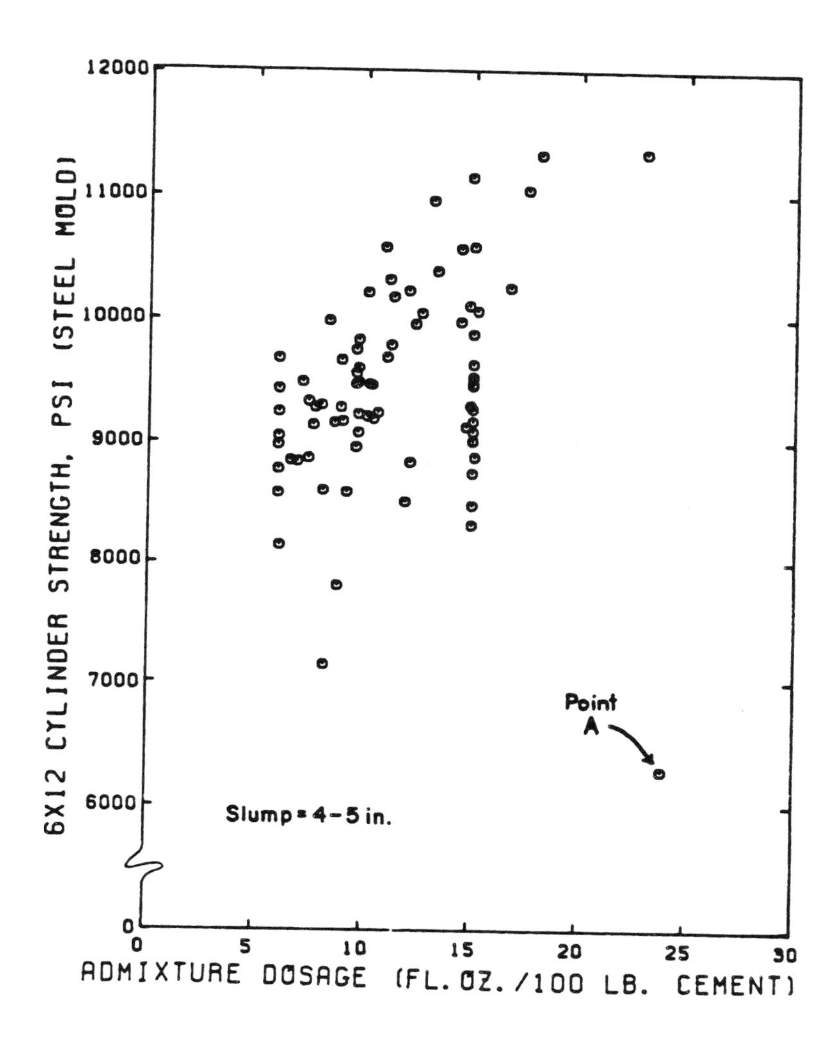

Fig. 4.17 Effect of superplasticizer dosage on the 56-day compressive strength of concrete.

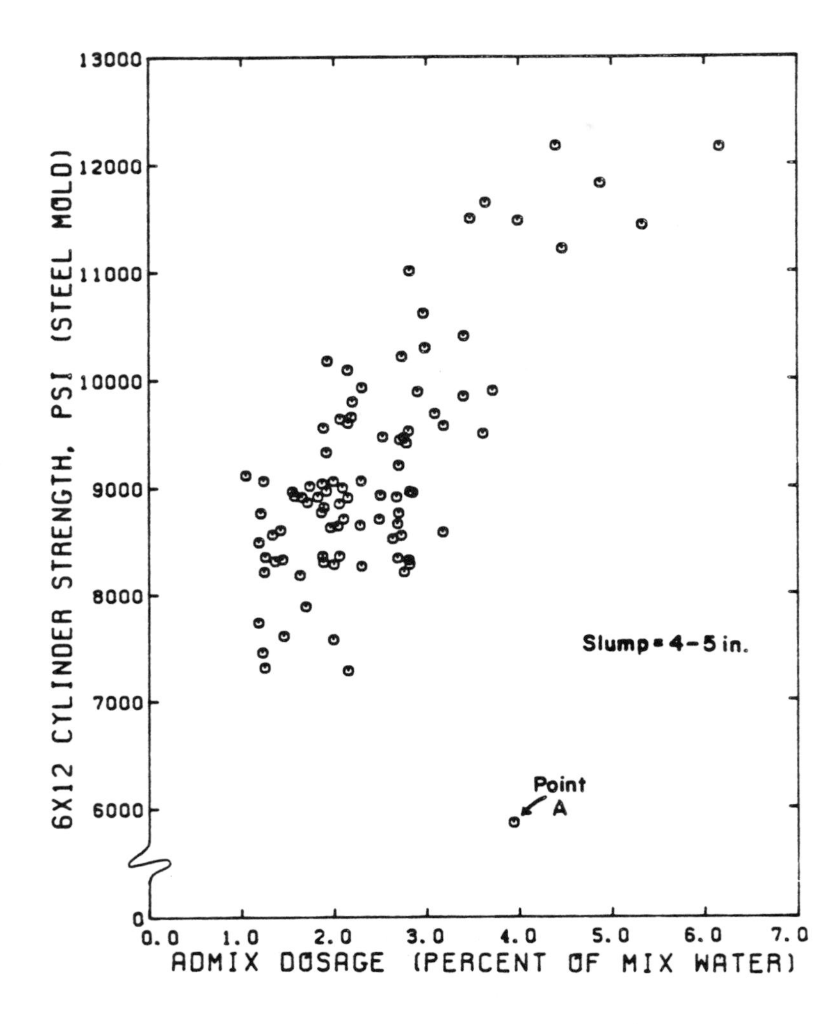

Fig. 4.18 Effect of superplasticizer dosage (expressed as a percent by weight of total mixing water) on the 28-day compressive strength of concrete for all mixes made containing superplasticizer.

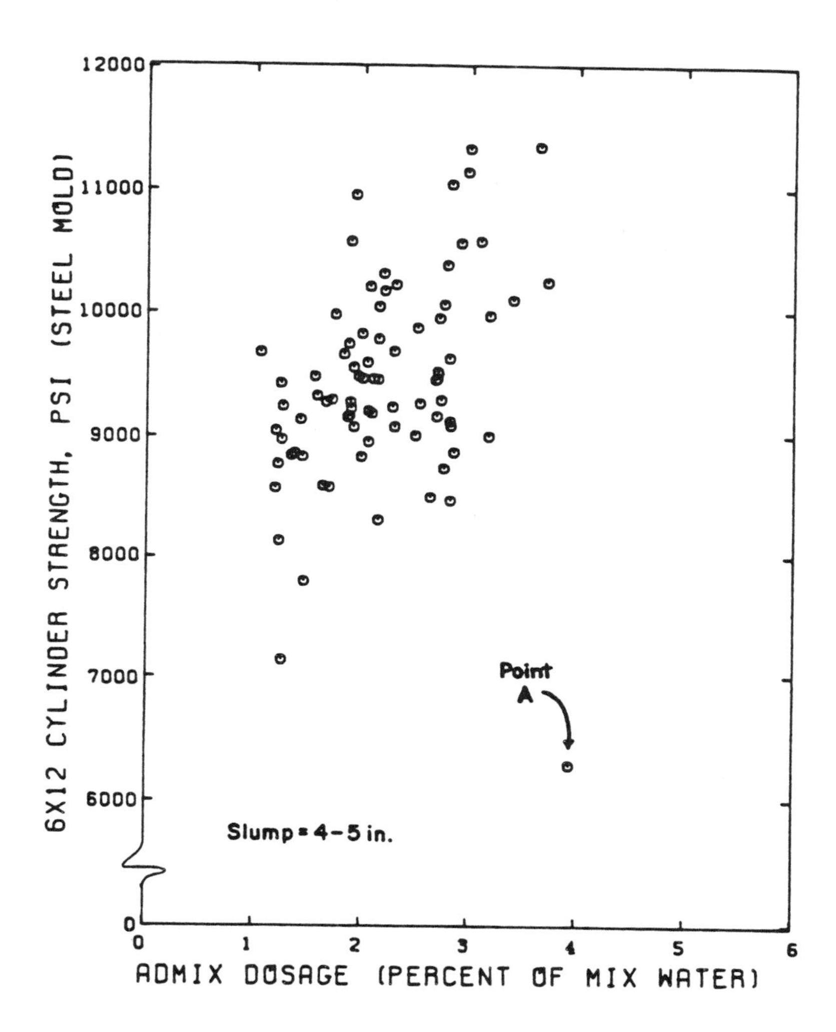

Fig. 4.19 Effect of superplasticizer dosage (expressed as a percent by weight of total mixing water) on the 56-day compressive strength of concrete.

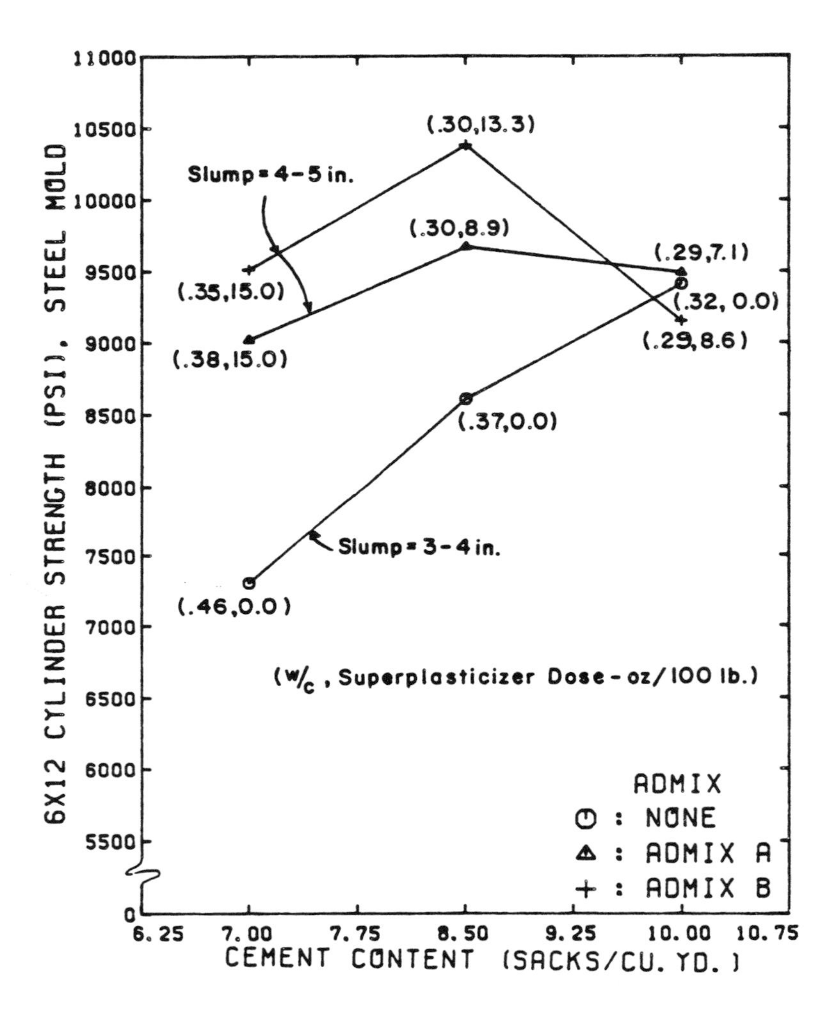

Fig. 4.20 Effect of superplasticizer and superplasticizer brand on the 56-day compressive strength of concrete for mixes having a CA/FA ratio of 2.0 and made with 7.0 to 10.0 sacks of type II cement per cu.yd., 1/2-in. limestone E, and sand C.

The strength increase in concrete compressive strength due to the addition of superplasticizer varied for different brands of superplasticizer for given mix proportions. However, based on the two superplasticizing admixtures used in this study, no consistent trend was found concerning the effect of superplasticizer brand on concrete compressive strength.

4.6 Coarse Aggregate Size

After the cement and both chemical and mineral admixtures, the coarse aggregate maximum size had the greatest influence on the compressive strength of high strength concrete. Three maximum sizes of crushed limestone coarse aggregate, 1/2-in., 3/4-in., and 1-in., were included in this study. The results of a comparison between gravel and crushed limestone coarse aggregates are presented in Section 4.7.

4.6.1 Cement Content. For concrete mixes containing no admixture, the compressive strength was highly dependent on the maximum size of coarse aggregate for cement contents ranging from 7 sacks/cu.yd. to 10 sacks/cu.yd., as shown in Figs. 4.21 and 4.22. For mixes containing 7 sacks/cu.yd., the effect of the maximum size of the coarse aggregate on concrete strength was directly related to the effect of that aggregate on the mixing water demand for a given workability. The 1-in. max. size coarse aggregate, having the smallest total surface area and consequently the lowest mixing water demand for a given slump, resulted in the highest compressive concrete strength regardless of test age and CA/FA ratio, for mixes containing 7 sacks/cu.yd. The 1/2-in. max. size

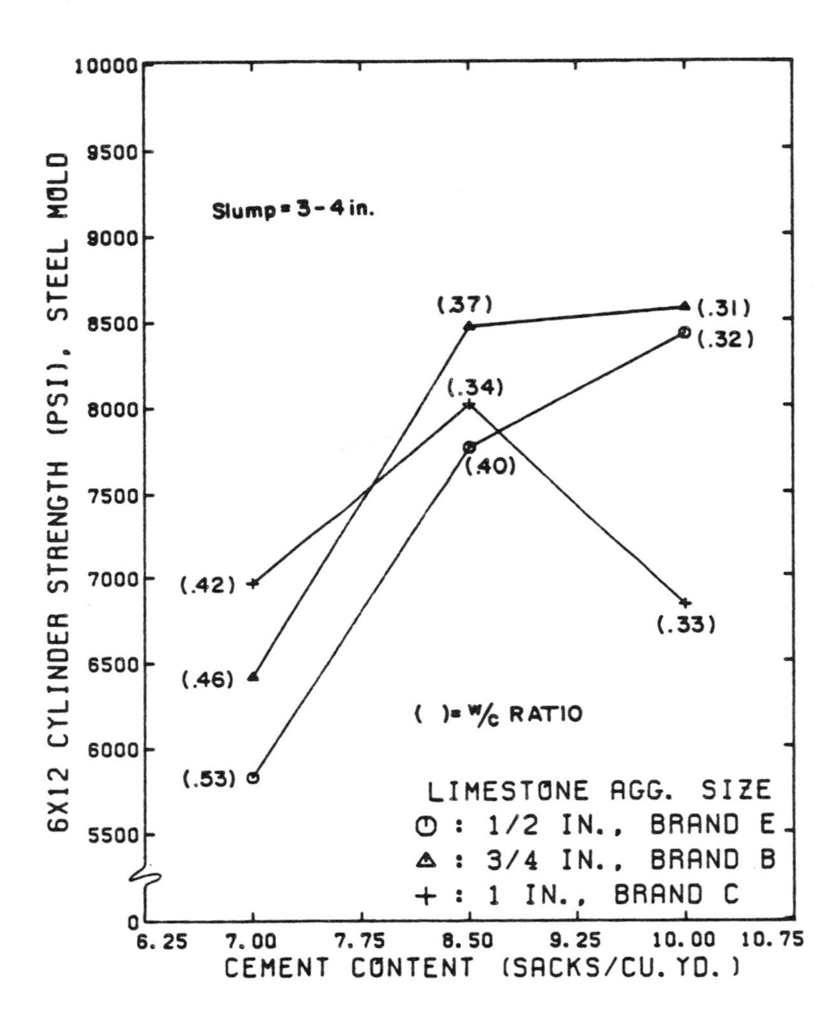

Fig. 4.21 Effect of coarse aggregate max. size and cement content on the 28-day compressive strength of concrete for mixes having a CA/FA ratio of 2.0 and made with type II cement, crushed limestone coarse aggregate, sand B, and no admixture.

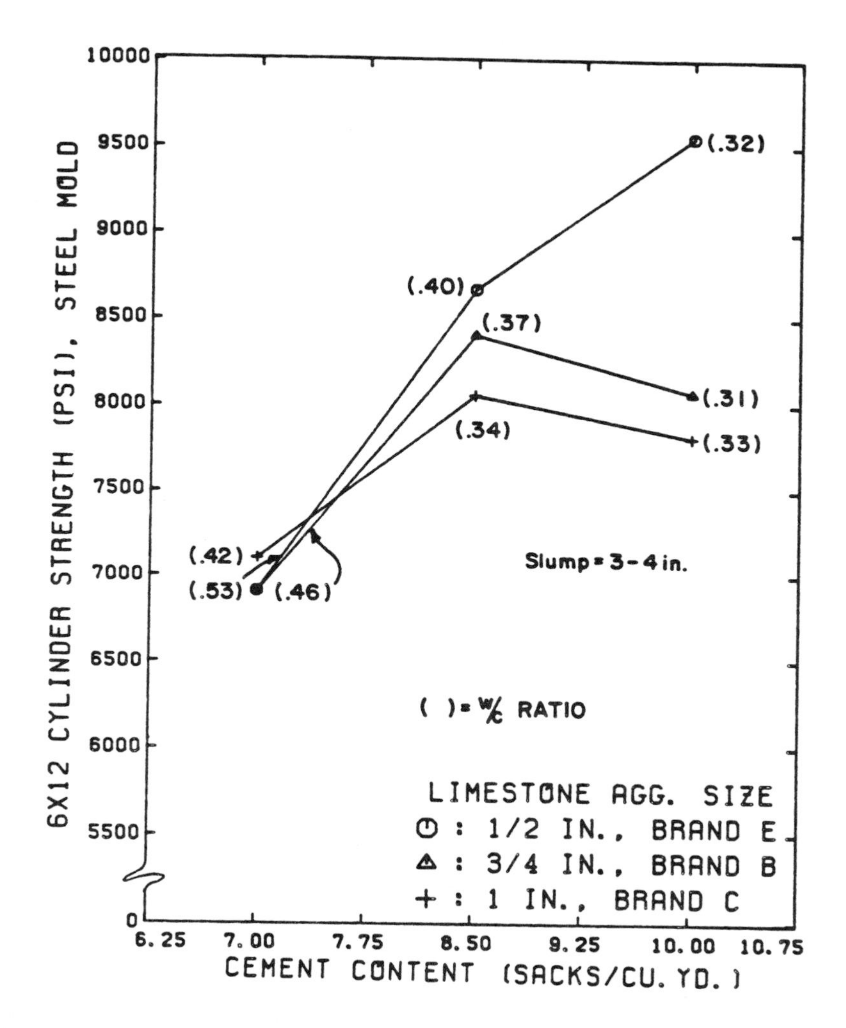

Fig. 4.22 Effect of coarse aggregate max. size and cement content on the 56-day compressive strength of concrete for mixes having a CA/FA ratio of 2.0 and made with type II cement, crushed limestone coarse aggregate, sand B, and no admixture.

coarse aggregate mixes produced the lowest compressive strengths for a cement content of 7 sacks/cu.yd.

For concrete mixes containing 8.5 sacks/cu.yd., the w/c ratio alone did not govern compressive strength. As shown in Fig. 4.21, the 3/4-in. aggregate produced the highest 28-day concrete strength for a cement content of 8.5 sacks/cu.yd., even though the w/c ratio was higher than for the mix containing 1-in. coarse aggregate. At 56 days, the 1/2-in. max. size aggregate mix containing 8.5 sacks/cu.yd. produced the greatest concrete compressive strength even though its water/cement ratio of 0.40 was the highest of the three mixes. In mixes having cement contents of 10 sacks/cu.yd., the 1/2-in. max. size coarse aggregate produced the highest concrete strength at 56 days. The compressive strength of several 10-sack mixes was less than that of some 8.5-sack mixes made with 3/4-in. aggregate and less than that of all batches containing 1-in. coarse aggregate.

As shown by Fig. 4.22, 10-sack concrete mixes containing 1/2-in. max. size coarse aggregate and no admixtures, and having a w/c ratio of approximately 0.32, achieved strengths in excess of 9500 psi at 56 days. A concrete strength of 9000 psi was also produced with a mix made using a 1-in. max. size crushed limestone coarse aggregate and a cement content of 10-sacks/cu.yd.

For mixes containing superplasticizer, 1/2-in. max. size aggregate was compared to 3/4-in. max. size aggregate. Variations in fineness modulus between shipments of sand from a single source hampered this analysis somewhat. However, for any combination of materials for which the fineness moduli were identical, the 1/2-in. aggregate concrete was stronger at 56 days than concrete made with 3/4-in. aggregate, for a 4- to 5-in. slump, as shown in Figs. 4.23 and 4.24 for a fineness modulus of 2.57. It should be noted that different admixture dosages were used for producing the same slump concrete for different mix proportions.

The difference between the effects of the two coarse aggregates should be most apparent in concrete mixes containing the most coarse aggregate, which in this study was for any mix with a coarse/fine aggregate ratio of 2.0. For a CA/FA ratio of 2.0, mixes containing 1/2-in. aggregate are stronger at 56 days for any cement content, independent of the w/c ratio and superplasticizer dosage, even though the 3/4-in. aggregate mixes contains less water and more admixture.

Concrete mixes made with 3/4-in. max. size coarse aggregate achieved strengths of approximately 9,000 psi at 56 days, while mixes made with 1/2-in. max. size aggregate achieved compressive strengths of 10,000 psi.

4.6.2 Coarse/Fine Aggregate Ratio. Trends can be seen in the compressive strength of concretes made with different maximum size coarse aggregates as a function of the CA/FA ratio.

Figures 4.25 through 4.28 show these relationships for mixes made with cement contents of 7 sacks/cu.yd. and 10 sacks/cu.yd., containing no admixture. Concrete made with 1-in. max. size coarse aggregates showed a reduction in strength with an increase in CA/FA for

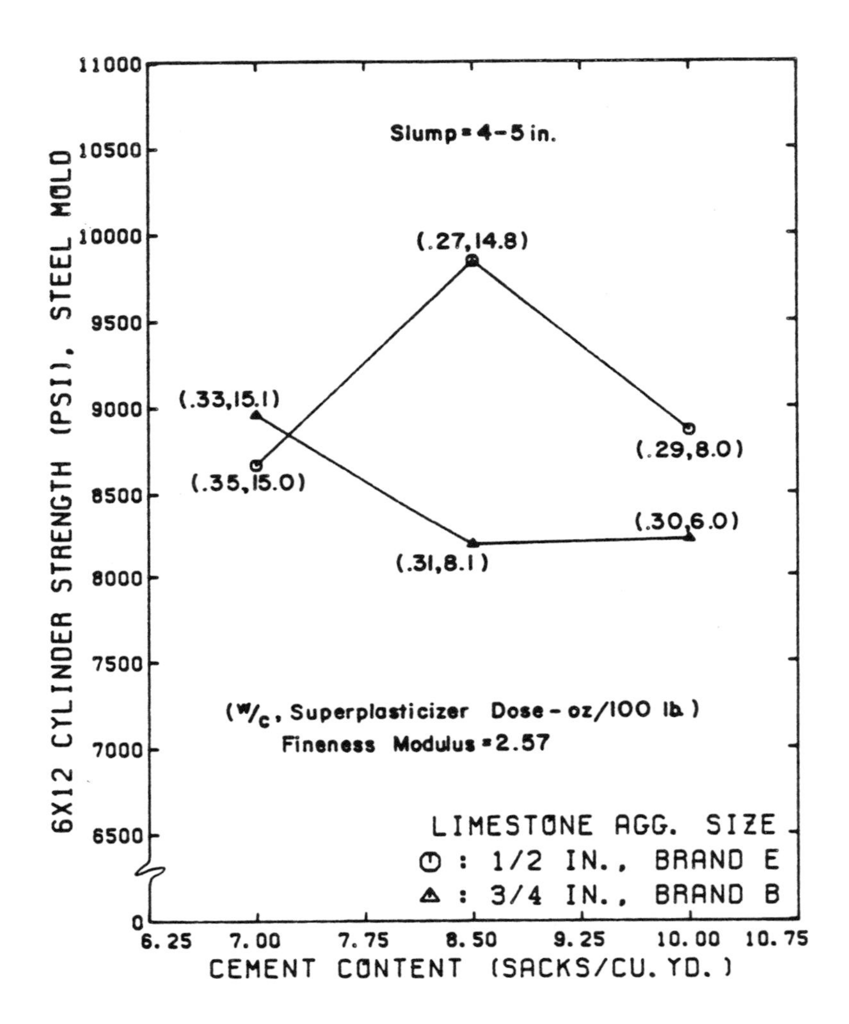

Fig. 4.23 Effect of coarse aggregate max. size and cement content on the 28-day compressive strength of concrete for mixes having a CA/FA ratio of 2.0 and made with type II cement, crushed limestone coarse aggregate, sand B, and superplasticizer B.

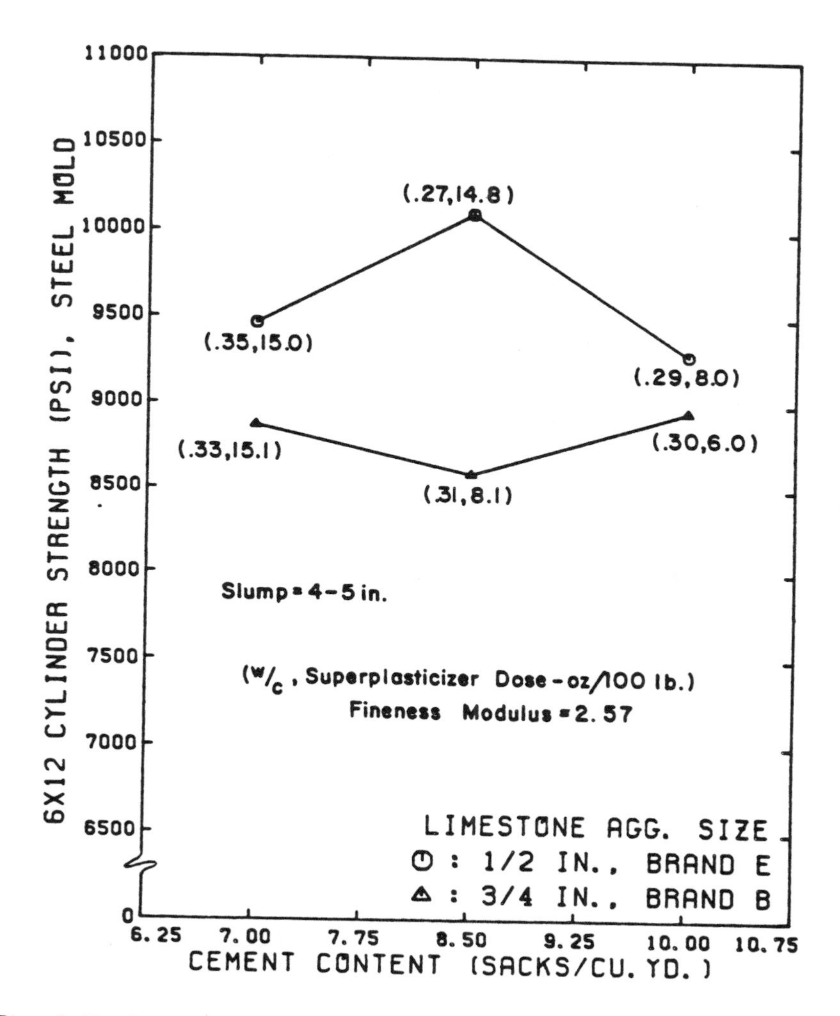

Fig. 4.24 Effect of coarse aggregate max. size and cement content on the 56-day compressive strength of concrete for mixes having a CA/FA ratio of 2.0 and made with type II cement, crushed limestone coarse aggregate, sand B, and superplasticizer B.

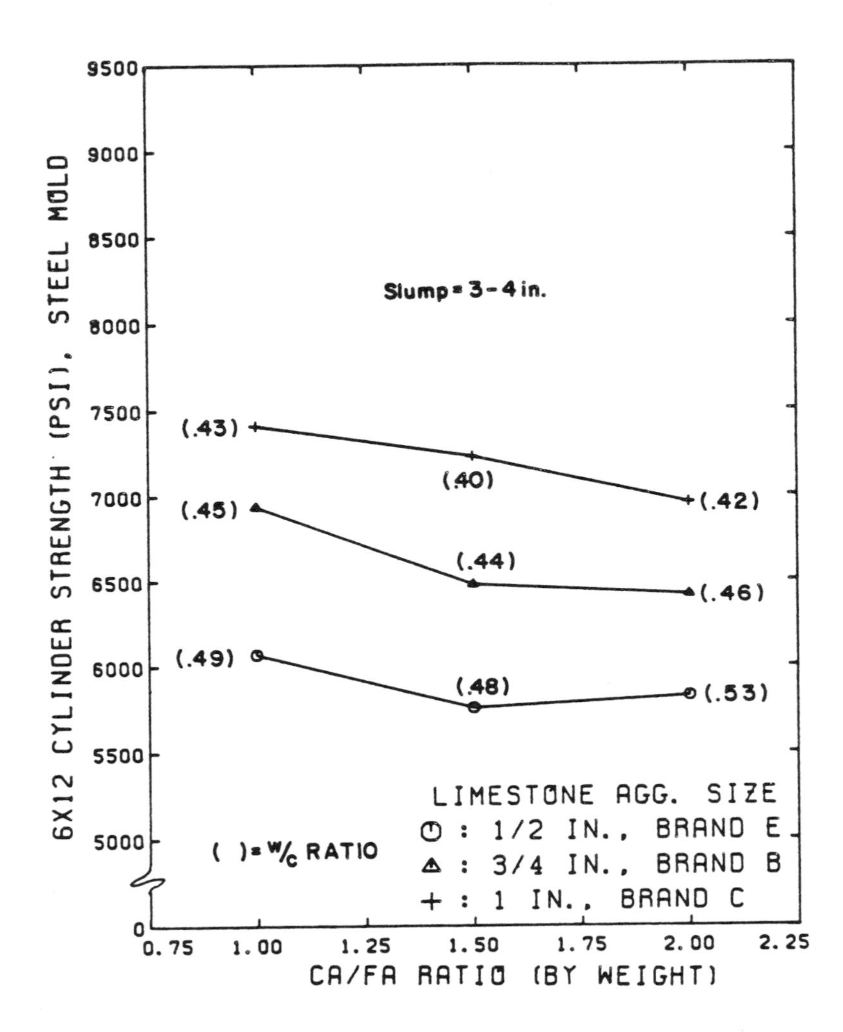

Fig. 4.25 Effect of coarse aggregate max. size and CA/FA ratio on the 28-day compressive strength of concrete for mixes having a cement content of 7.0 sacks/cu.yd. and made with type II cement, crushed limestone coarse aggregate, sand B, and no admixture.

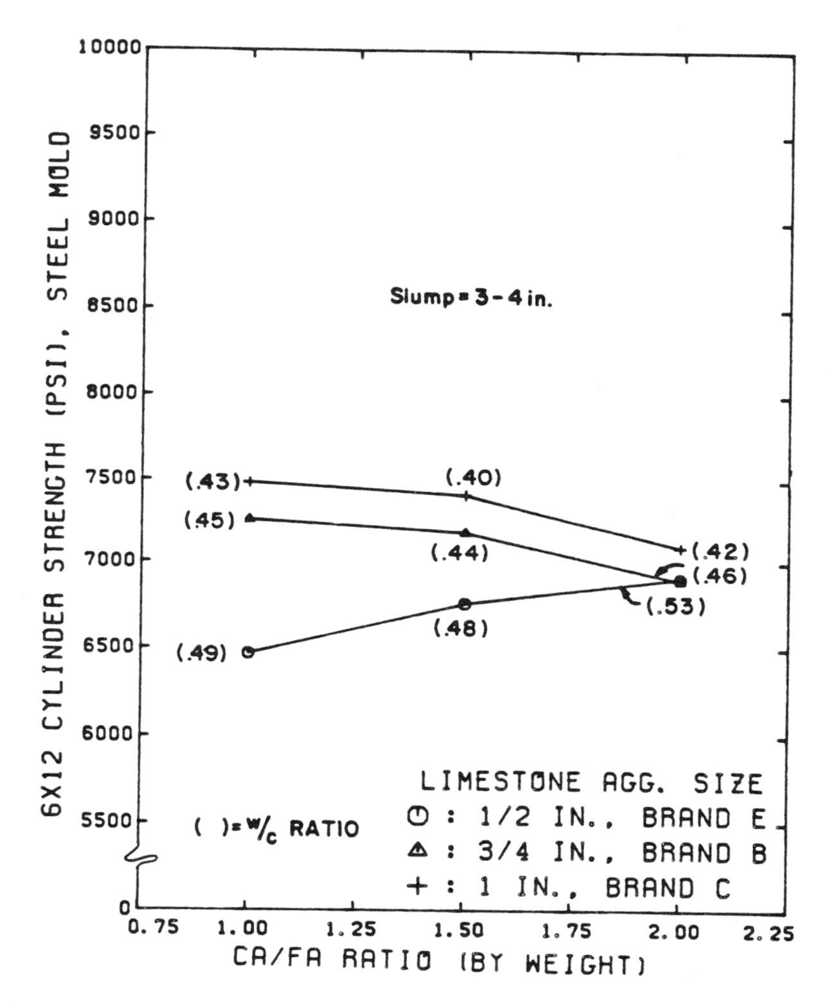

Fig. 4.26 Effect of coarse aggregate max. size and CA/FA ratio on the 56-day compressive strength of concrete for mixes having a cement content of 7.0 sacks/cu.yd. and made with type II cement, crushed limestone coarse aggregate, sand B, and no admixture.

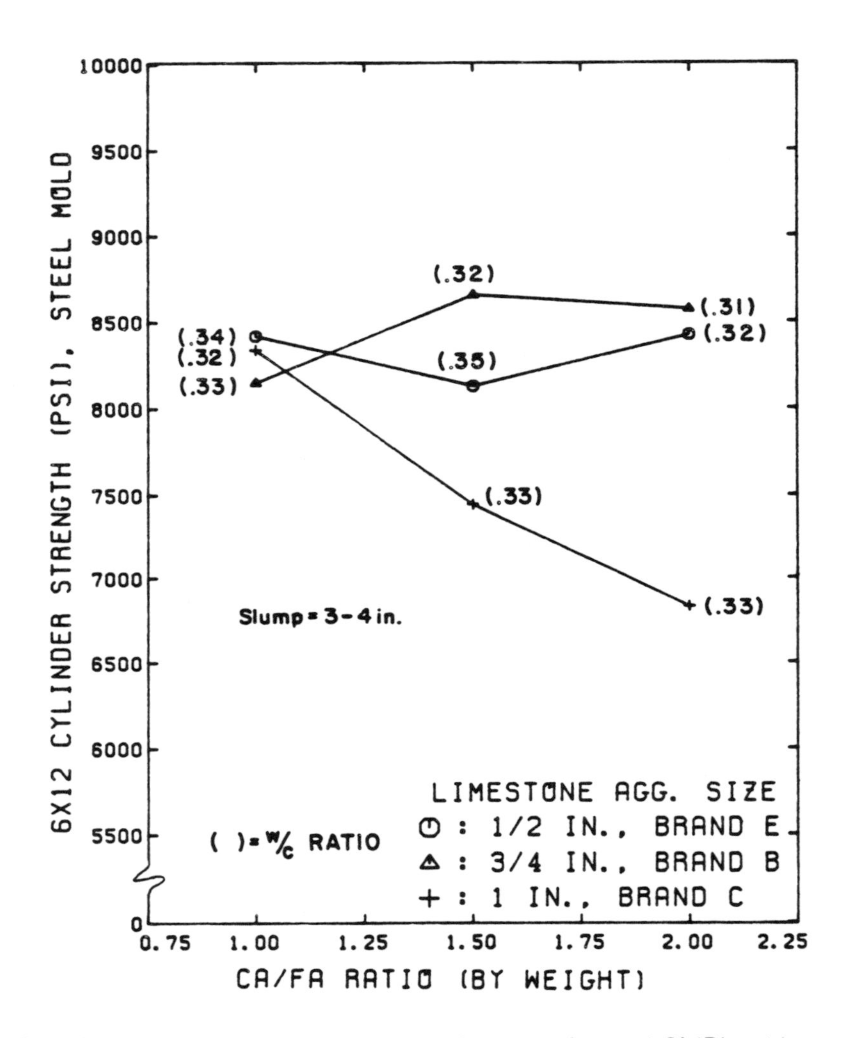

Fig. 4.27 Effect of coarse aggregate max. size and CA/FA ratio on the 28-day compressive strength of concrete for mixes having a cement content of 10 sacks/cu.yd. and made with type II cement, crushed limestone coarse aggregate, sand B, and no admixture.

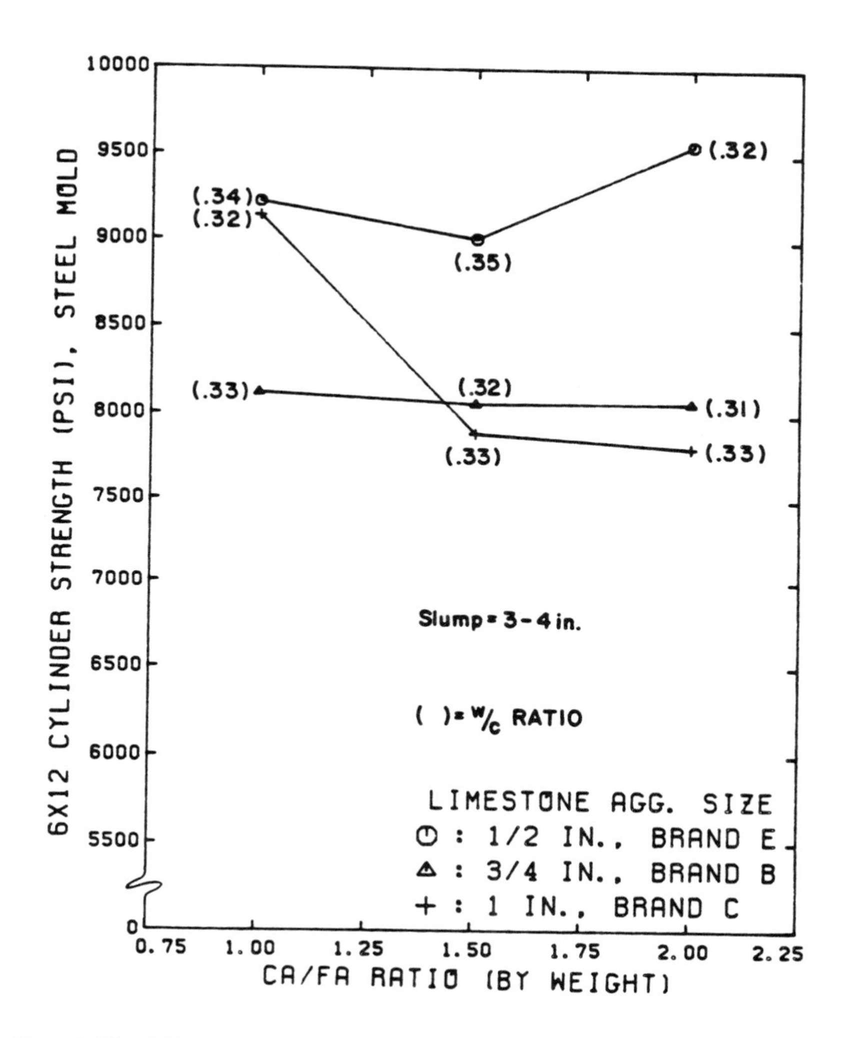

Fig. 4.28 Effect of coarse aggregate max. size and CA/FA ratio on the 56-day compressive strength of concrete for mixes having a cement content of 10 sacks/cu.yd. and made with type II cement, crushed limestone coarse aggregate, sand B, and no admixture.

any cement content. This relationship does not correspond to the trend in concrete strength predicted based on the w/c ratio of these mixes.

For mixes made with 3/4-in. maximum size aggregate, compressive strength decreased with an increase in CA/FA for a cement content of 7.0 sacks/cu.yd., as shown in Figs. 4.25 and 4.26. At higher cement contents, the compressive strength of mixes containing 3/4-in. max. size aggregate remained unchanged or increased with an increase in CA/FA, as seen in Figs. 4.27 and 4.28.

The 1/2-in. aggregate concretes tended to increase in compressive strength at 56 days with an increase in CA/FA ratio, as seen in Figs. 4.26 and 4.28, regardless of cement content. An exception to this, though, was concrete made with 1/2-in. aggregate D, which had a low bulk specific gravity and unit weight. Concrete made with aggregate D had a lower compressive strength with an increase in CA/FA, as shown by Fig. 4.29.

Mixes containing superplasticizer did not exhibit a clear trend in compressive strength as a function of CA/FA, since variations in admix dosage for a given slump appeared to control the concrete strength. Figures 4.30 through 4.33 show the effects of coarse aggregate size and CA/FA ratio on the compressive strength of concrete mixes containing 7, 8.5, and 10 sacks/cu.yd. and superplasticizer.

In general, for CA/FA ratios of over 1.5, concretes made with 1/2-in. max. size aggregates showed higher compressive strengths than concretes made using 3/4-in. max. size aggregate for all cement contents.

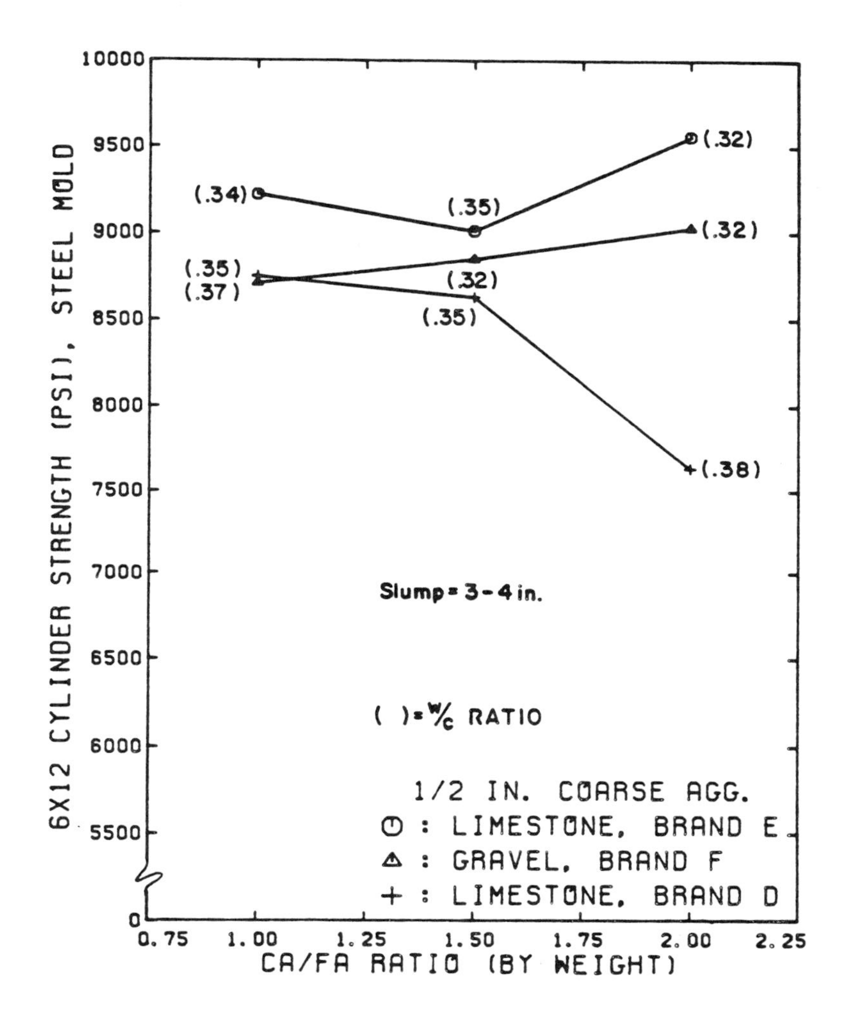

Fig. 4.29 Effect of coarse aggregate type and CA/FA ratio on the 56-day compressive strength of concrete for mixes having a cement content of 10 sacks/cu.yd. and made with type II cement, crushed limestone coarse aggregate, sand B, and no admixture.

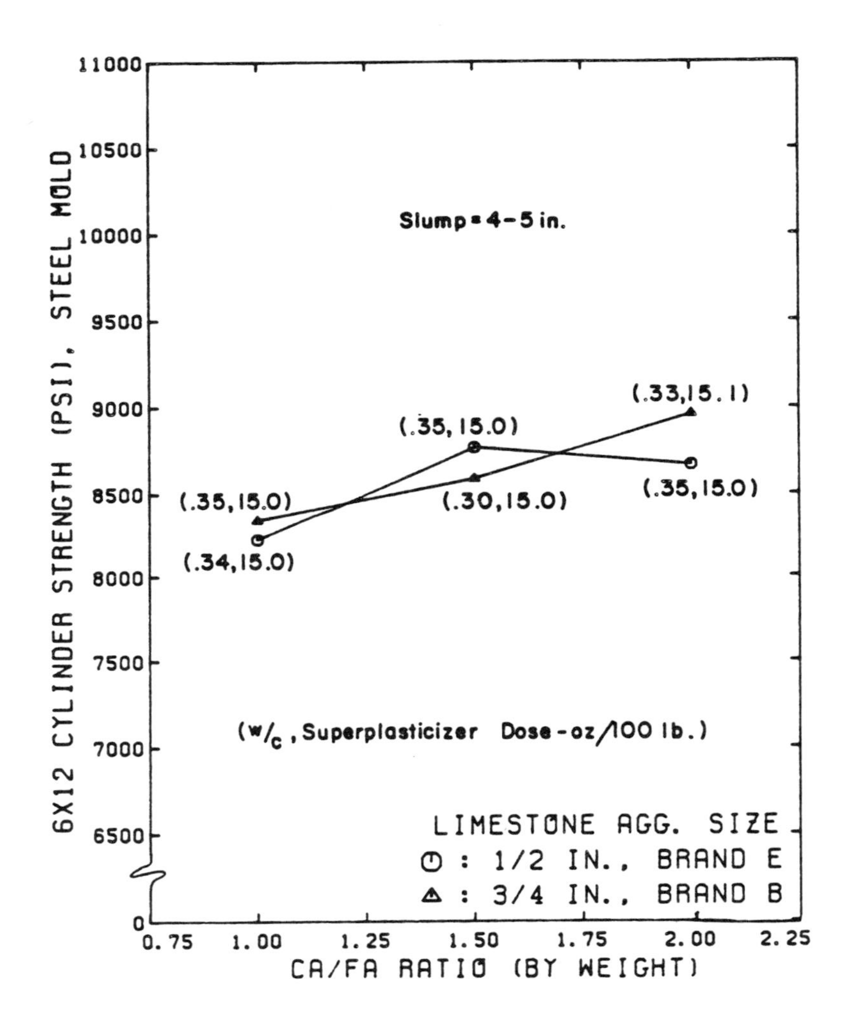

Fig. 4.30 Effect of coarse aggregate max. size and CA/FA ratio on the 28-day compressive strength of concrete for mixes having a cement content of 7.0 sacks/cu.yd. and made with type II cement, crushed limestone coarse aggregate, sand B, and superplasticizer B.

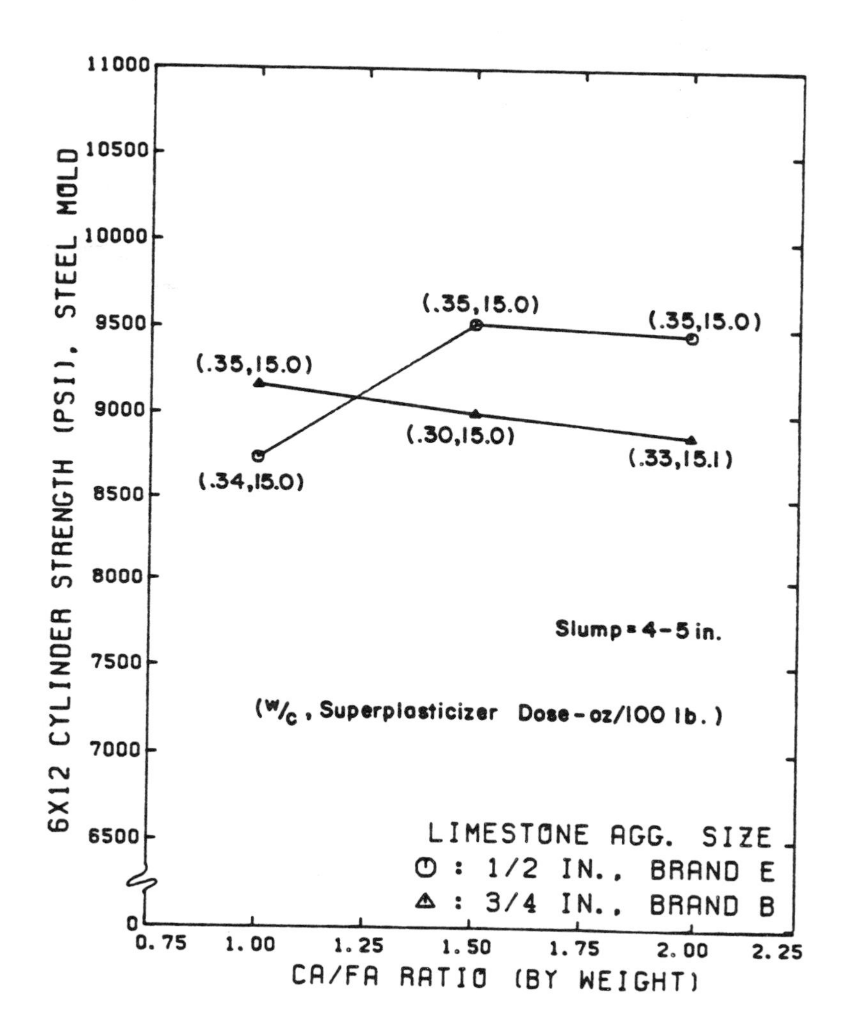

Fig. 4.31 Effect of coarse aggregate max. size and CA/FA ratio on the 56-day compressive strength of concrete for mixes having a cement content of 7.0 sacks/cu.yd. and made with type II cement, crushed limestone coarse aggregate, sand B, and superplasticizer B.

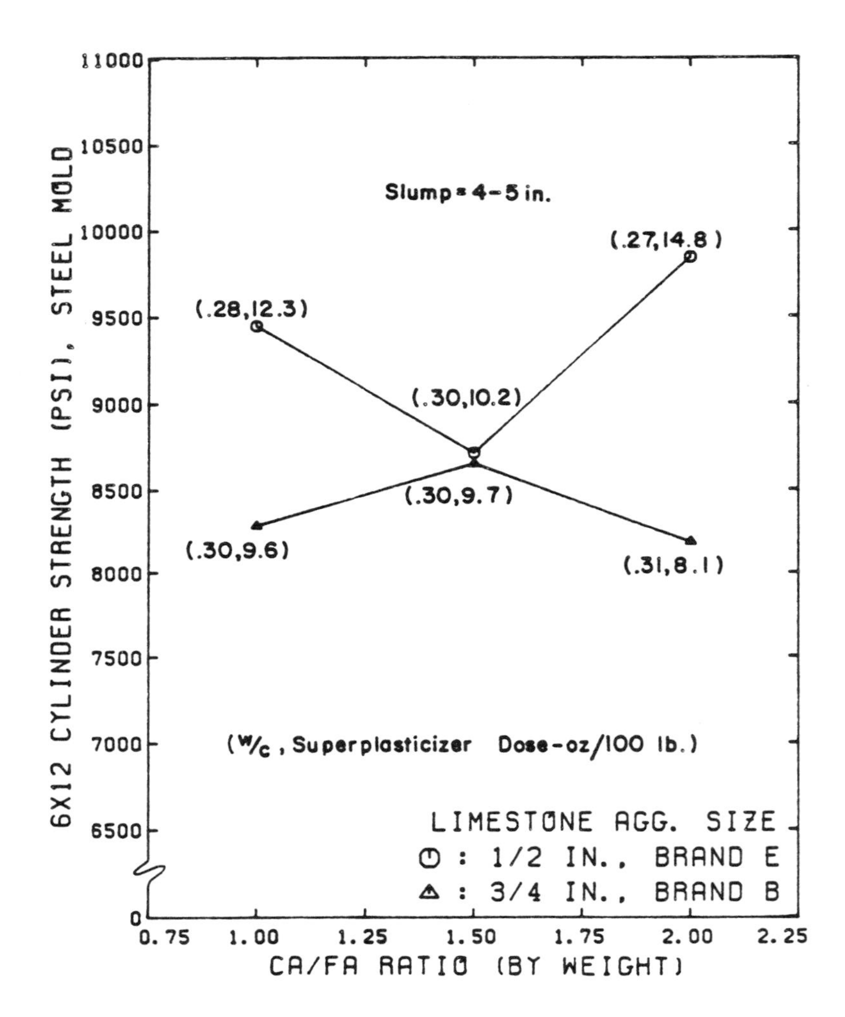

Fig. 4.32 Effect of coarse aggregate max. size and CA/FA ratio on the 28-day compressive strength of concrete for mixes having a cement content of 8.5 sacks/cu.yd. and made with type II cement, crushed limestone coarse aggregate, sand B, and superplasticizer B.

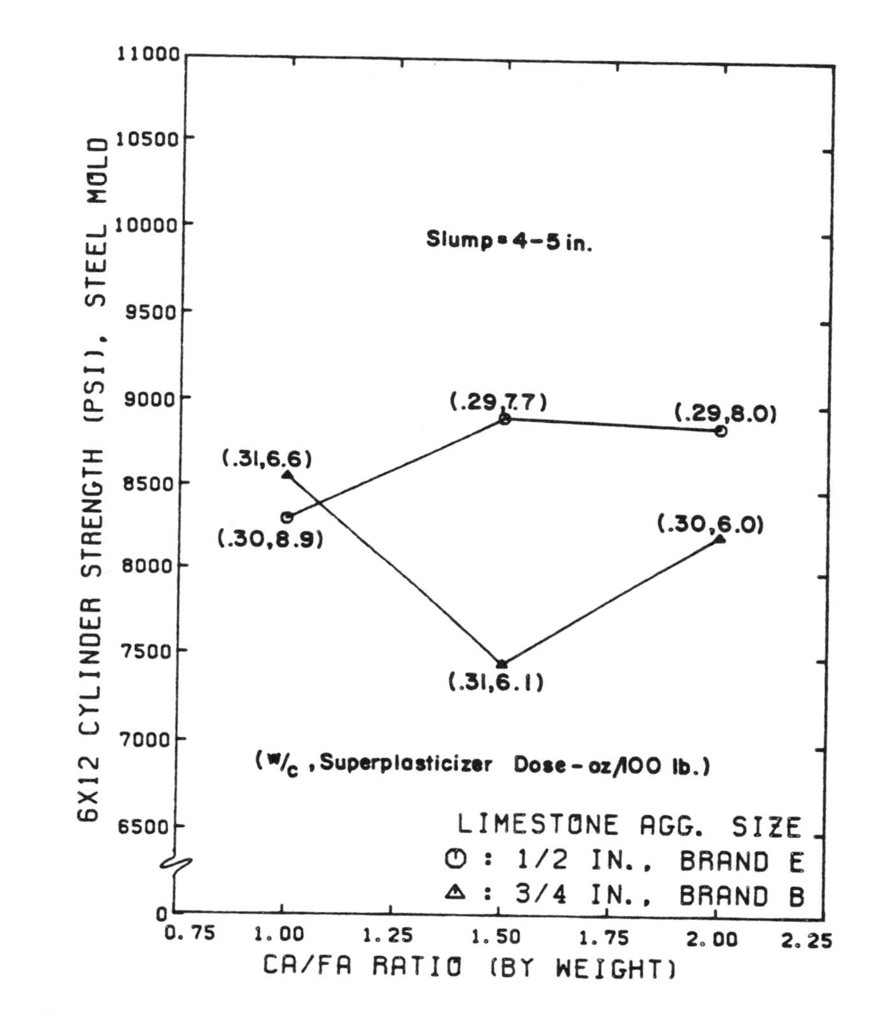

Fig. 4.33 Effect of coarse aggregate max. size and CA/FA ratio on the 28-day compressive strength of concrete for mixes having a cement content of 10 sacks/cu.yd. and made with type II cement, crushed limestone coarse aggregate, sand B, and superplasticizer B.

4.6.3 Specimen Age. Mixes containing no superplasticizer gained compressive strength from 28 to 56 days of age at greater rates for smaller sizes of coarse aggregate and greater w/c ratios, as shown in Figs. 4.34 through 4.36. For mixes containing superplasticizer, the lower the w/c ratio, the larger the strength gain from 28 to 56 days, as shown in Fig. 4.37. These relationships in strength gain with time were typical of nearly all concrete batches tested in this study.

A summary of the effects of aggregate maximum size and specimen age on compressive strength of concrete is presented in Table 4.1.

4.7 Coarse Aggregate Gradation

All but three of the mixes in this study were made with "as received" coarse aggregates. The gradations of all aggregates used are shown in Appendix A.

To compare concretes made with coarse aggregates having different gradations, aggregate B was separated by sieve size and recombined into three predetermined size distributions. As shown in Fig. 4.38, these size distributions correspond to coarse, medium and fine gradations within the limits on aggregate gradation for use in concrete according to *Texas 1982 Standard Specifications for Construction of Highways, Streets and Bridges*, Item 421.2(3), and the ASTM Standard Specification for Concrete Aggregates, C33-80.

One concrete mix containing the same dosage of superplasticizer was made using each of the three coarse aggregate gradations shown in Fig. 4.38. The only variable besides coarse aggregate gradation in

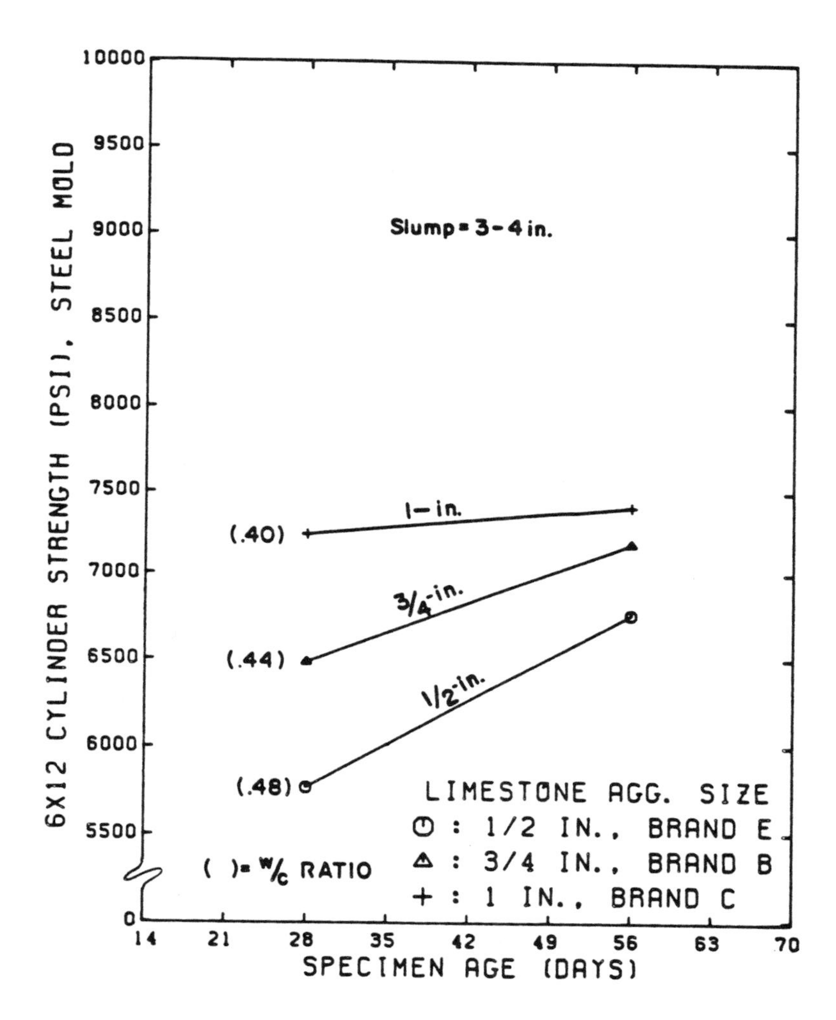

Fig. 4.34 Effect of coarse aggregate max. size and specimen age on the compressive strength of concrete for mixes having a cement content of 7 sacks/cu.yd. and a CA/FA ratio of 1.5 and made with type II cement, crushed limestone coarse aggregate, sand B, and no admixture.

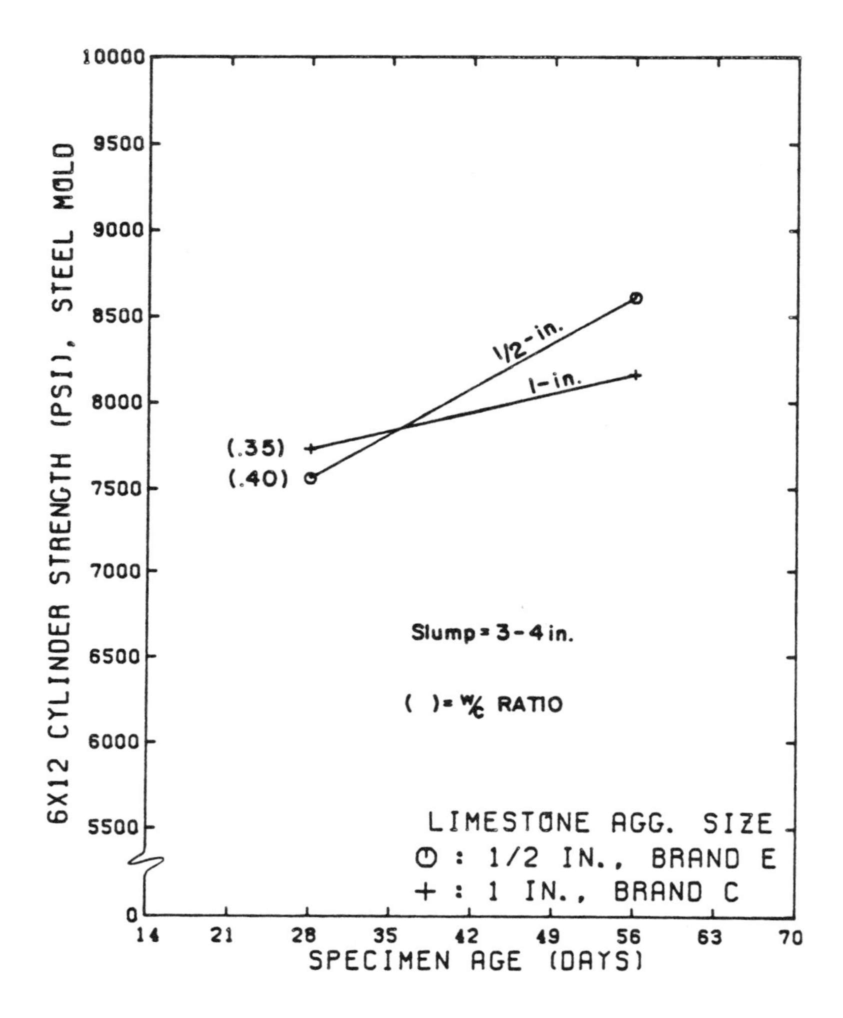

Fig. 4.35 Effect of coarse aggregate max. size and specimen age on the compressive strength of concrete for mixes having a cement content of 8.5 sacks/cu.yd. and a CA/FA ratio of 1.5 and made with type II cement, crushed limestone coarse aggregate, sand B, and no admixture.

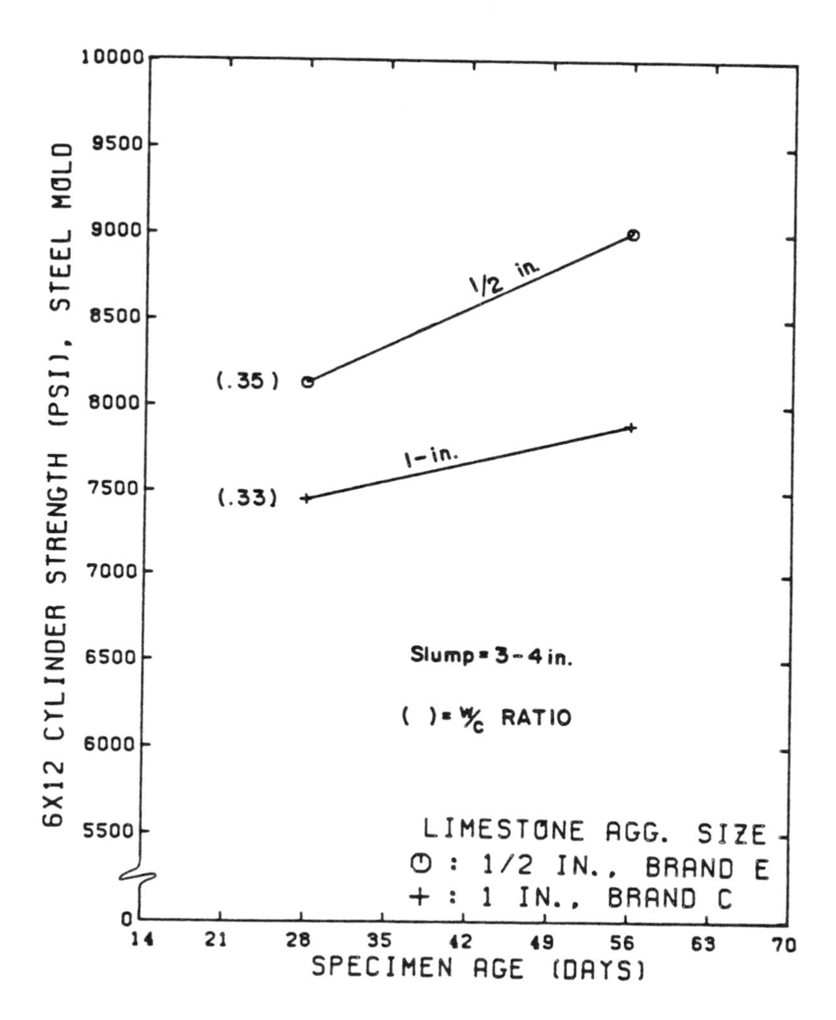

Fig. 4.36 Effect of coarse aggregate max. size and specimen age on the compressive strength of concrete for mixes having a cement content of 10 sacks/cu.yd. and a CA/FA ratio of 1.5 and made with type II cement, crushed limestone coarse aggregate, sand B, and no admixture.

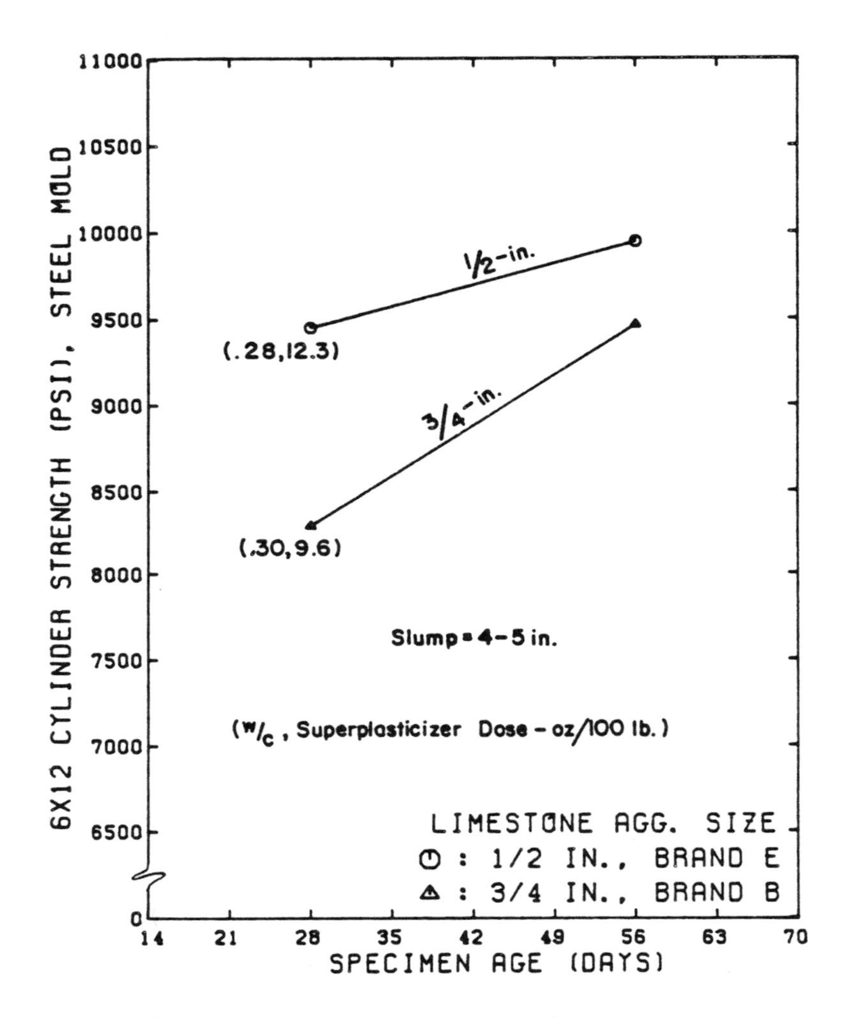

Fig. 4.37 Effect of coarse aggregate max. size and specimen age on the compressive strength of concrete for mixes having a cement content of 8.5 sacks/cu.yd. and a CA/FA ratio of 1.0 and made with type II cement, crushed limestone coarse aggregate, sand B, and superplasticizer B.

TABLE 4.1 Comparison of the Average Rate of Increase in Compressive Strength of Concrete from a Test Age of 28 Days to 56 Days for Mixes made using Different Sizes of Crushed Limestone Coarse Aggregate (includes No Mixes which Contain Fly Ash).

Crushed Limestone Coarse Aggregate	Gain in Compressive Strength of Concrete from 28 to 56 Days	
	With No Admixture	With Superplasticizer
1/2-in. Aggregate E	13.0%	6.7%
3/4-in. Aggregate B	6.5%	7.2%
1-in. Aggregate C	4.3%	---

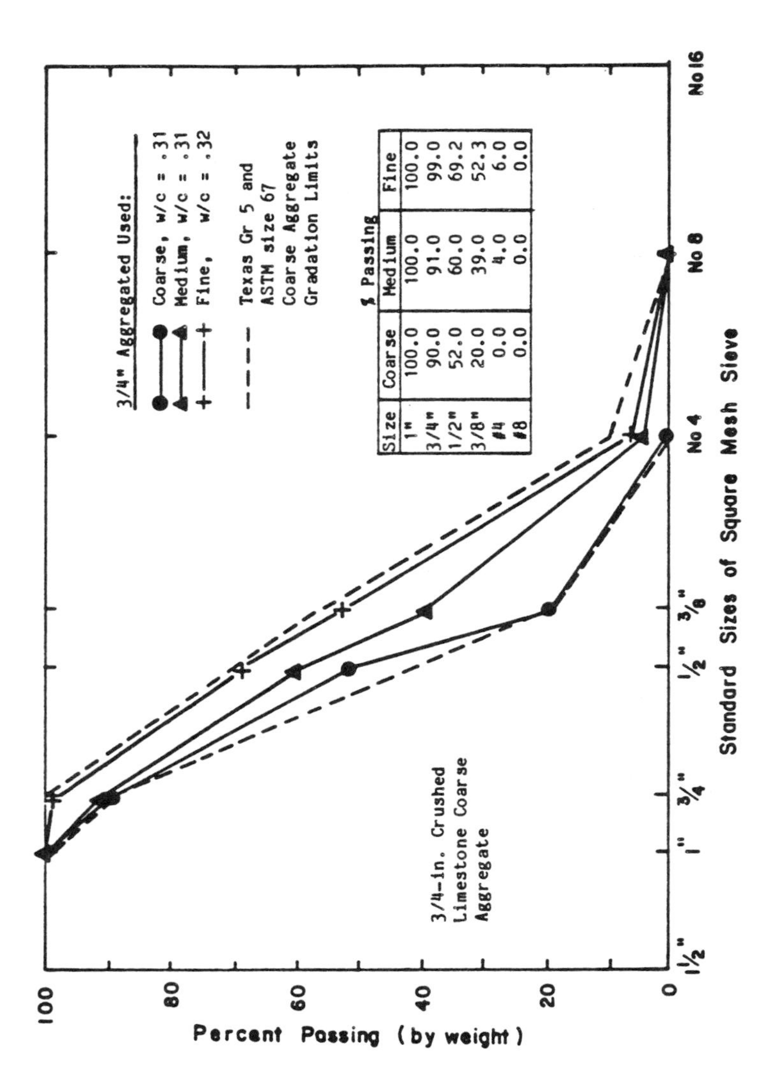

Size	Coarse	Medium	Fine
1"	100.0	100.0	100.0
3/4"	90.0	91.0	99.0
1/2"	52.0	60.0	69.2
3/8"	20.0	39.0	52.3
#4	0.0	4.0	6.0
#8	0.0	0.0	0.0

Fig. 4.38 Coarse aggregate gradation curves for three aggregates used in concrete mixes made to study the effect of coarse aggregate gradation on compressive strength of concrete.

these mixes was the mixing water requirement for producing a 4-in. slump.

The results of the compressive strength tests performed on these three mixes are shown in Fig. 4.39. At 56 days, the compressive strength increases as the w/c ratio of the concrete mix decreases, as shown in Fig. 4.39. The mix made with the coarsest coarse aggregate gradation required the least amount of mix water resulting in the highest compressive strength. Concretes made with the fine gradation of coarse aggregate resulted in the highest mixing water demand and therefore the lowest compressive strength at 56 days.

4.8 Coarse Aggregate Type

Two types of coarse aggregate were used: crushed limestone and natural gravel. In addition, limestone coarse aggregates taken from two different sources were considered. The purpose was to study how texture, shape, and mineralogy affect the compressive strength of high strength concrete.

4.8.1 Cement Content.

Two limestone coarse aggregates and one natural gravel were used in similar concrete mixes, with and without superplasticizer. Figures 4.40 and 4.41 show the effect of aggregate type on concrete compressive strength as a function of cement content at 28 days and 56 days of age for mixes containing no superplasticizer.

In general, using a crushed limestone having a dry rodded unit weight of 85 lb/cu.ft. and a bulk specific gravity of 2.46 (SSD) resulted in a higher mixing water demand and lower concrete strength for all cement contents at any test age than when using a crushed limestone

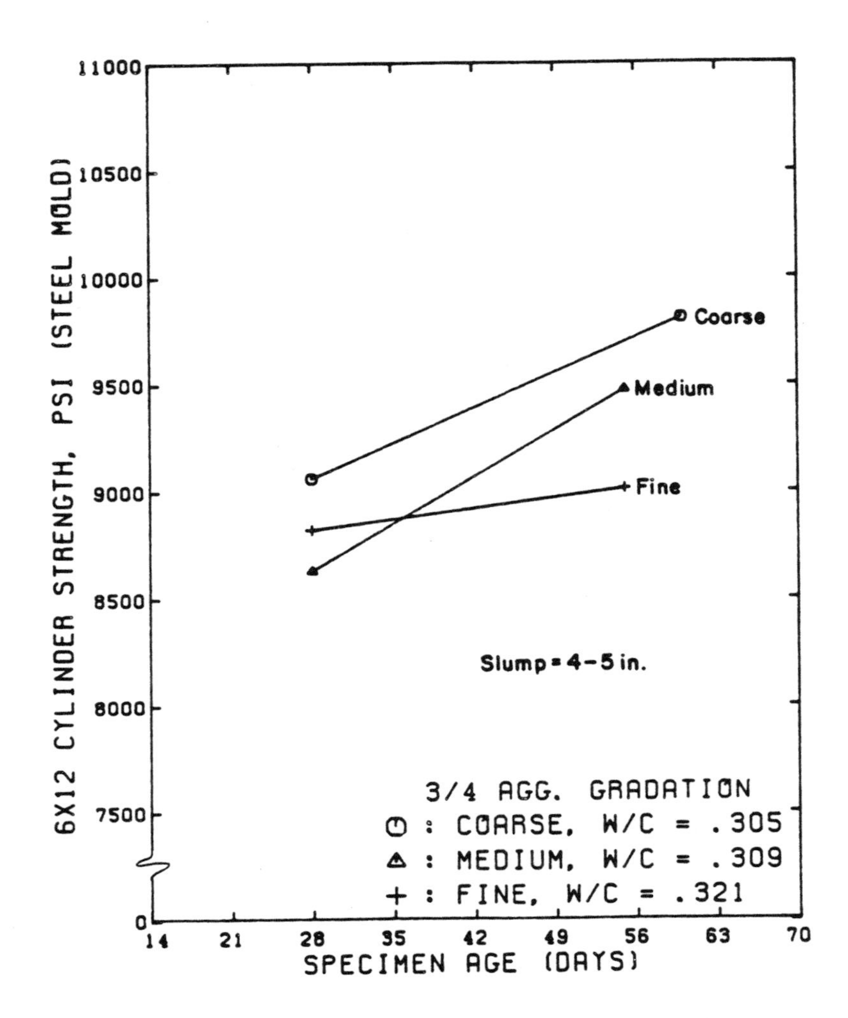

Fig. 4.39 Effect of coarse aggregate gradation and specimen age on the compressive strength of concrete for mixes having a cement content of 8.5 sacks/cu.yd. and a CA/FA ratio of 1.5 and made with type II cement, 3/4-in. crushed limestone coarse aggregate, sand B, and superplasticizer B.

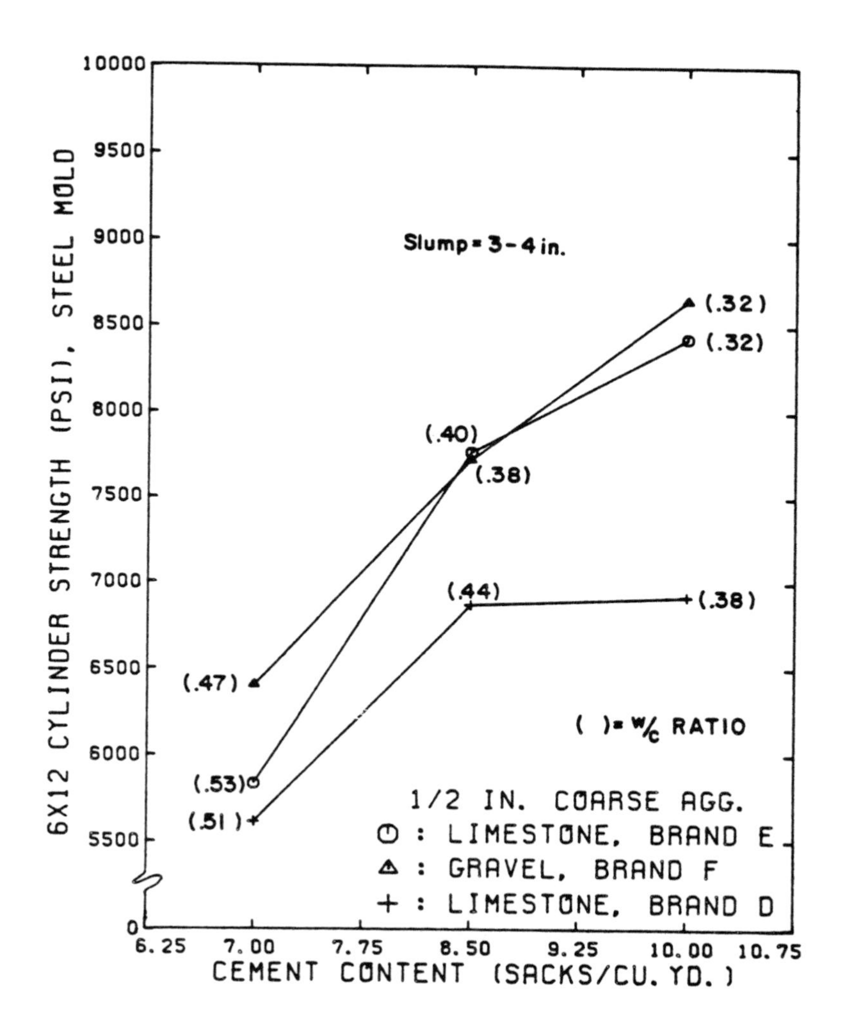

Fig. 4.40 Effect of coarse aggregate type and cement content on the 28-day compressive strength of concrete for mixes having a CA/FA ratio of 2.0 and made with type II cement, 1/2-in. max. size coarse aggregate, sand B, and no admixture.

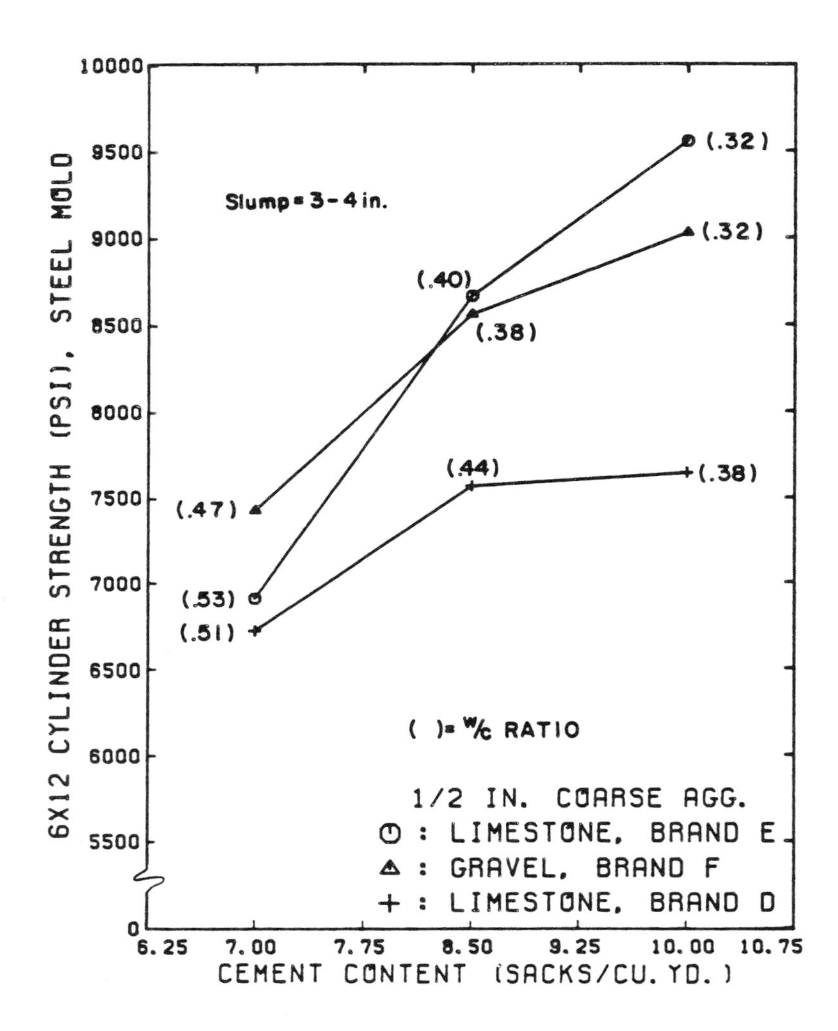

Fig. 4.41 Effect of coarse aggregate type and cement content on the 56-day compressive strength of concrete for mixes having a CA/FA ratio of 2.0 and made with type II cement, 1/2-in. max. size coarse aggregate, sand B, and no admixture.

of the same max. size but having a dry rodded unit weight of 95 lb/cu.ft. and bulk specific gravity of 2.65. The compressive strength difference between concretes made with the two limestones was greater for higher cement contents. For 7-sack mixes, the difference in strength was approximately 5%, but in 10-sack mixes the difference in strength was about 20%, at 28 or 56 days.

In addition, there was a difference in the optimum cement content above which no increase in strength was obtained from increasing the cement content of the mix for the two crushed limestones. The optimum cement content was higher for the more dense limestone aggregate.

Comparing the gravel mixes to the mixes made using the more dense limestone, the gravel concrete had a significantly lower water requirement at lower cement contents and a higher compressive strength, at 28 or 56 days. For mixes containing 8.5 sacks of cement/cu.yd., the difference in water requirement was small, resulting in similar concrete strengths for gravel and limestone concretes at 28 and 56 days. However, limestone concretes had a higher compressive strength than gravel concretes at 56 days in mixes containing 10 sacks of cement/cu.yd., even though the w/c ratio of both concretes was the same. Whereas gravel mixes achieved strengths of approximately 9,000 psi at 56 days with 10 sacks of cement/cu.yd., limestone mixes exceeded 9,500 psi, especially for a CA/FA ratio of 2.0.

For mixes containing superplasticizers there was no clear trend between aggregate type and cement content. Figures 4.42 through

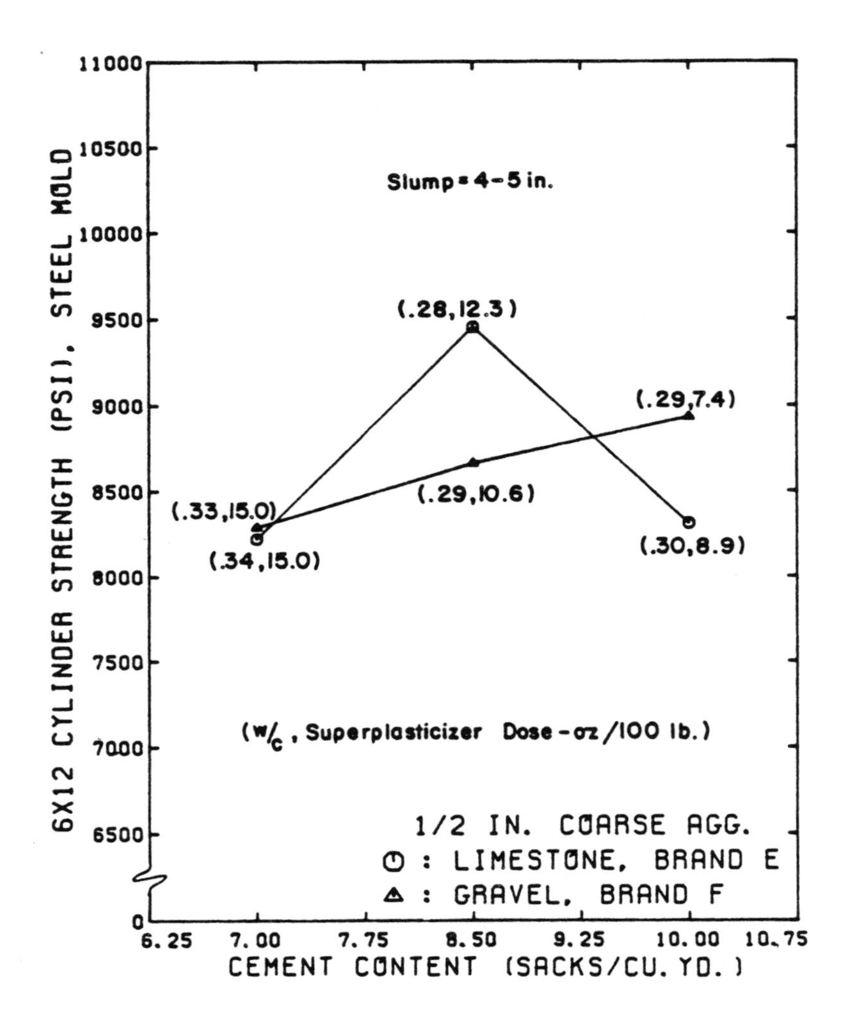

Fig. 4.42 Effect of coarse aggregate type and cement content on the 28-day compressive strength of concrete for mixes having a CA/FA ratio of 1.0 and made with type II cement, 1/2-in. max. size coarse aggregate, sand B, and superplasticizer B.

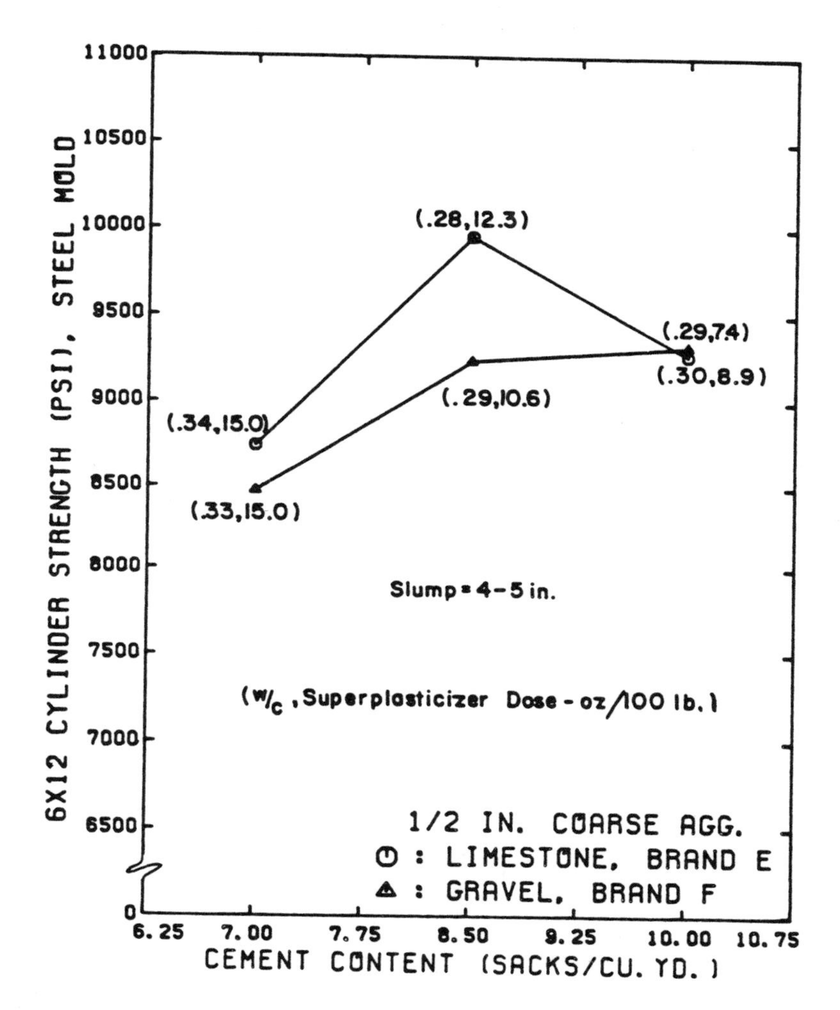

Fig. 4.43 Effect of coarse aggregate type and cement content on the 56-day compressive strength of concrete for mixes having a CA/FA ratio of 1.0 and made with type II cement, 1/2-in. max. size coarse aggregate, sand B, and superplasticizer B.

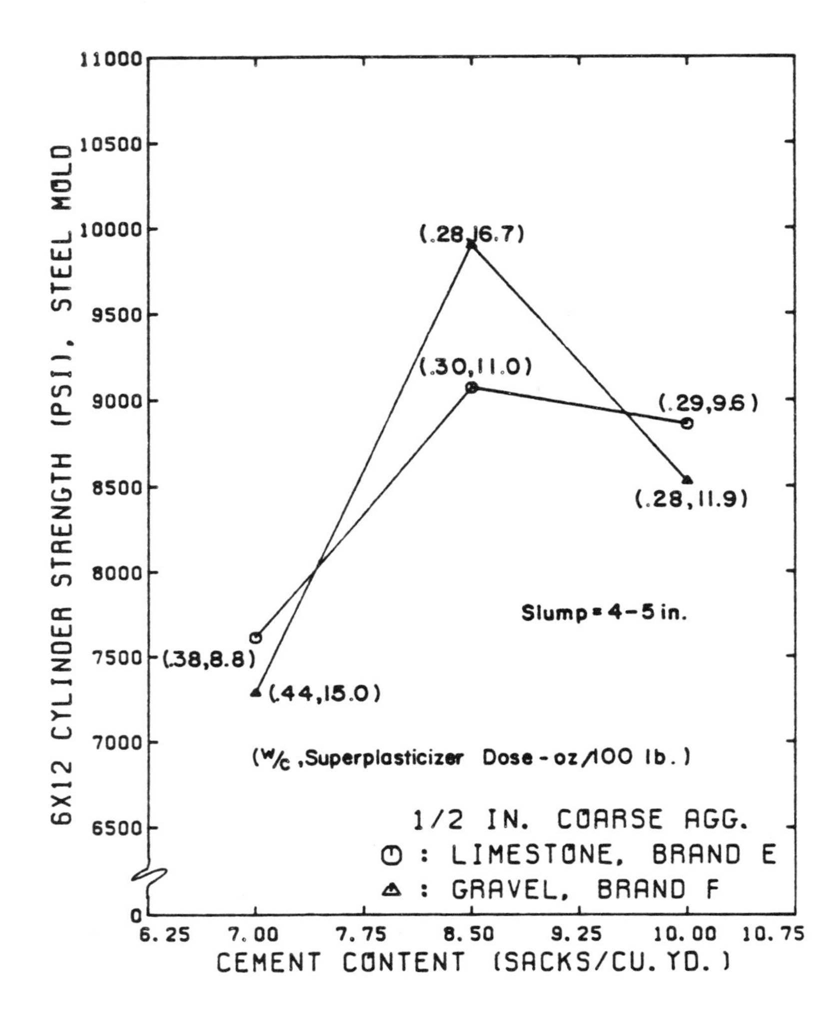

Fig. 4.44 Effect of coarse aggregate type and cement content on the 28-day compressive strength of concrete for mixes having a CA/FA ratio of 1.5 and made with type II cement, 1/2-in. max. size coarse aggregate, sand C, and superplasticizer B.

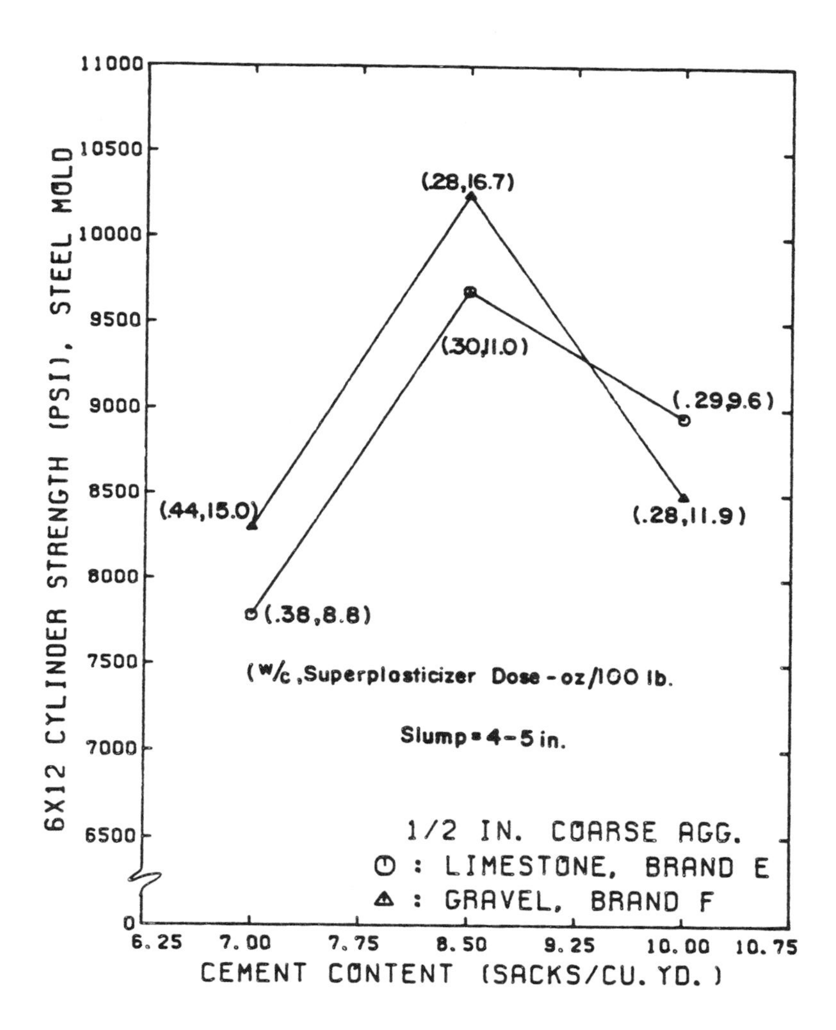

Fig. 4.45 Effect of coarse aggregate type and cement content on the 56-day compressive strength of concrete for mixes having a CA/FA ratio of 1.5 and made with type II cement, 1/2-in. max. size coarse aggregate, sand C, and superplasticizer B.

4.45 show compressive strength at 28 days and 56 days plotted versus cement content for concretes containing two different sands. Admixture dose affected the results considerably, but it can be seen from these figures that 10,000 psi compressive strengths can be achieved using gravel or crushed stone if a superplasticizer is added to the mix, regardless of CA/FA ratios.

4.8.2 Coarse/Fine Aggregate Ratio. In high cement content concrete mixes containing no admixtures, the general trend for gravel F and limestone E (DRUW = 95 lb/cu.ft., BSG = 2.65) concretes was for an increase in compressive strength with an increase in CA/FA. This same relationship was observed even for two concretes having the same w/c ratio but different CA/FA ratios. Increased strength with increased CA/FA was not as significant and less noticeable at 28 days than at 56 days of age.

For high cement content mixes containing limestone "D" (DRUW = 85 lb/cu.ft., BSG = 2.46) lower compressive strengths were obtained as the amount of coarse aggregate increased. This is shown in Fig. 4.46 for concrete mixes having a cement content of 8.5 sacks/cu.yd. Mixes containing no admixtures and having a cement factor of 7.0 sacks/cu.yd. showed no relationship betwee compressive strength and CA/FA ratio.

For gravel concrete containing superplasticizer, the compressive strength increases with higher CA/FA ratios at low cement contents as seen in Figs. 4.47 and 4.48. For higher cement contents, the compressive strength of gravel concretes tended to decrease for a CA/FA ratio of 2.0, as shown in Figs. 4.49 and 4.50. Limestone mixes containing

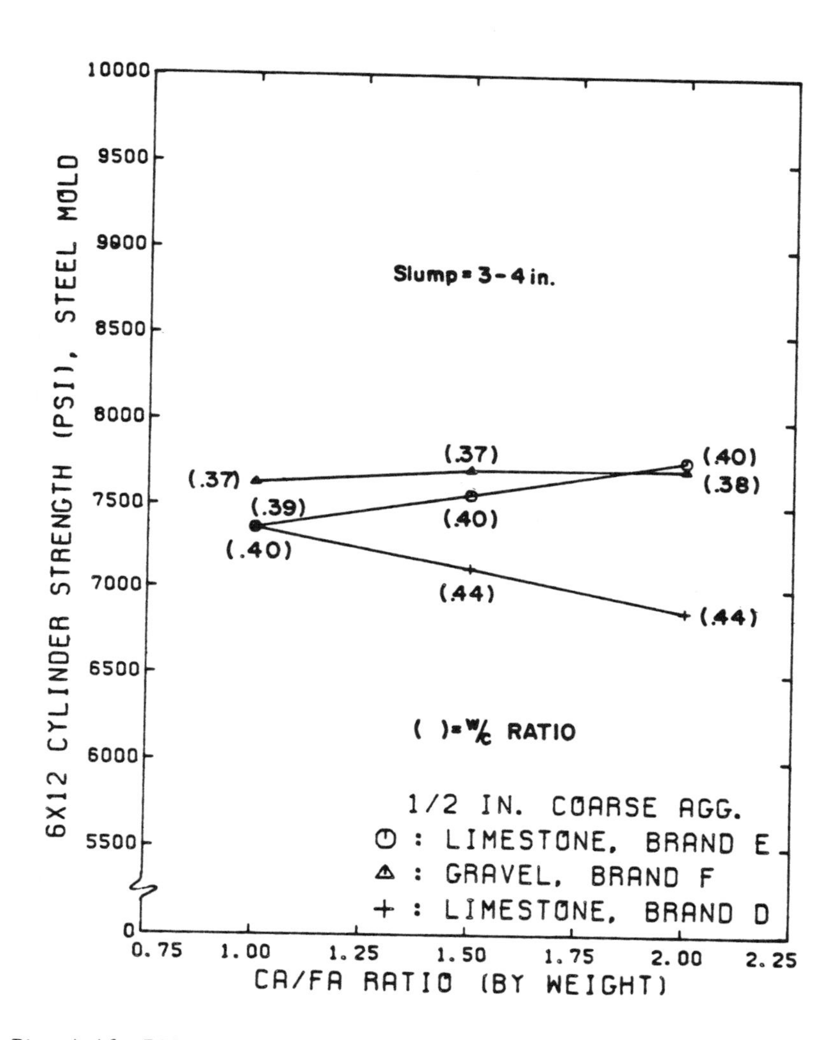

Fig. 4.46 Effect of coarse aggregate type and CA/FA ratio on the 28-day compressive strength of concrete for mixes having a cement content of 8.5 sacks/cu.yd. and made with type II cement, 1/2-in. max. size coarse aggregate, sand B, and no admixture.

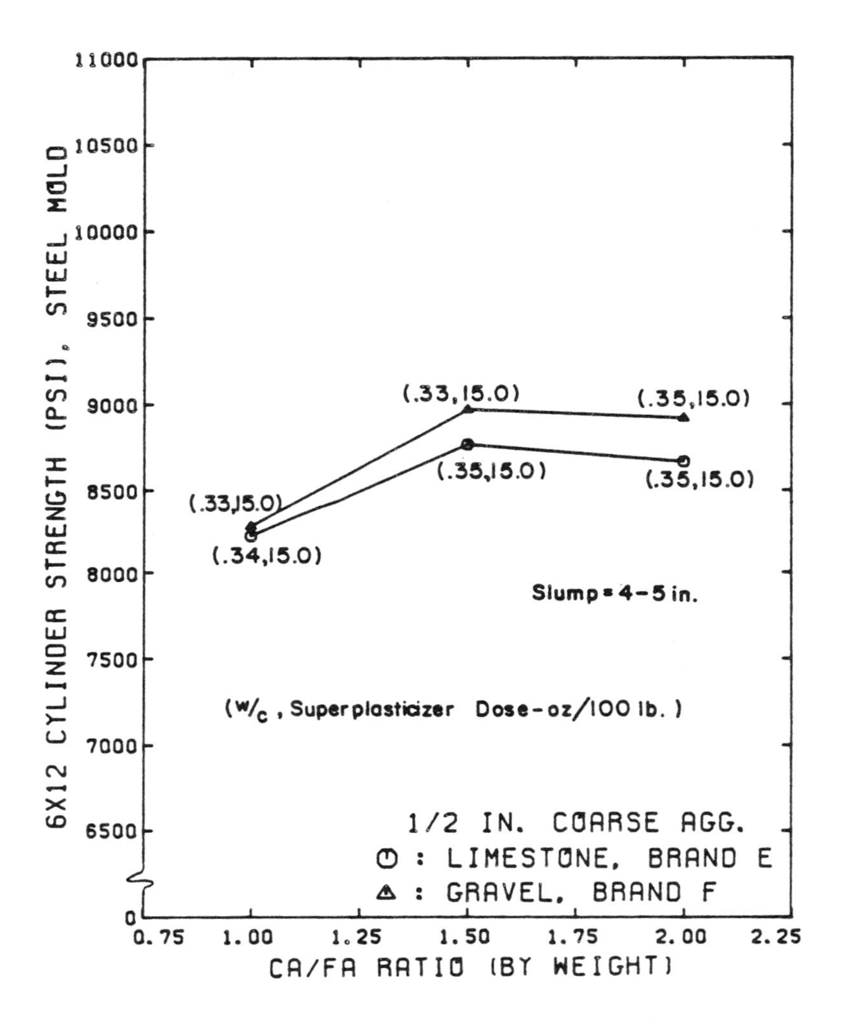

Fig. 4.47 Effect of coarse aggregate type and CA/FA ratio on the 28-day compressive strength of concrete for mixes having a cement content of 7.0 sacks/cu.yd. and made with type II cement, 1/2-in. max. size coarse aggregate, sand B, and superplasticizer B.

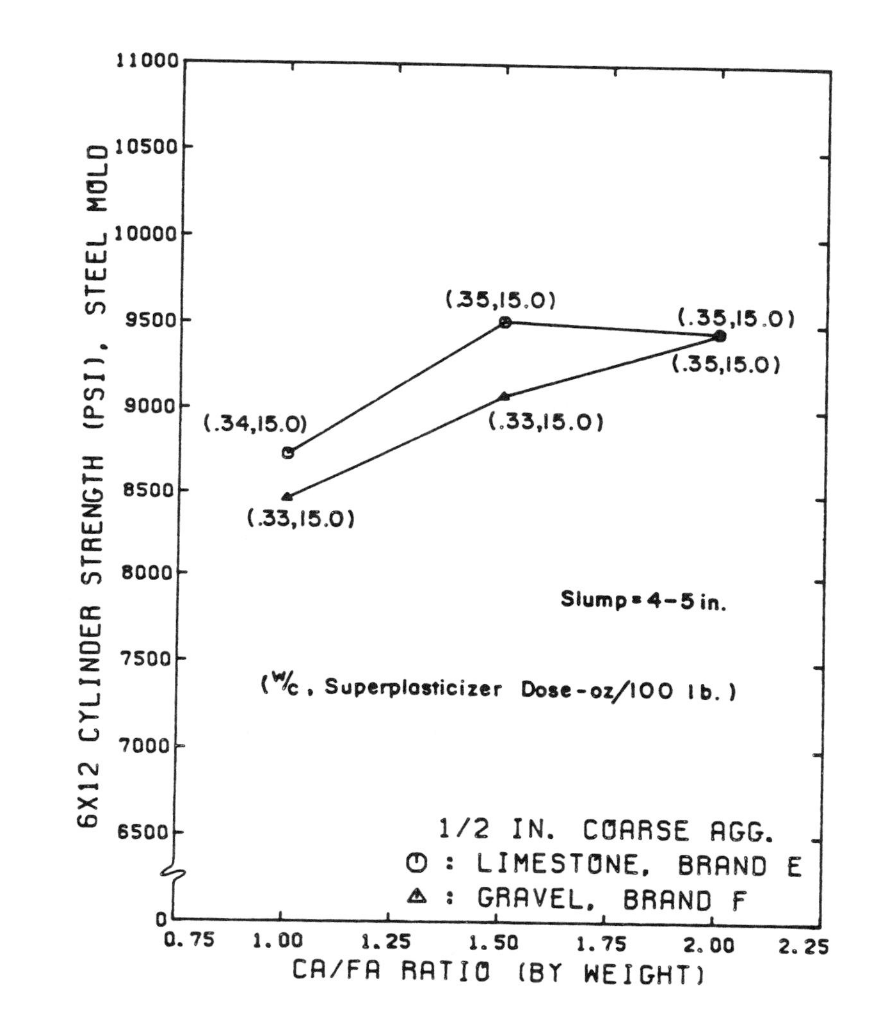

Fig. 4.48 Effect of coarse aggregate type and CA/FA ratio on the 56-day compressive strength of concrete for mixes having a cement content of 7.0 sacks/cu.yd. and made with type II cement, 1/2-in. max. size coarse aggregate, sand B, and superplasticizer B.

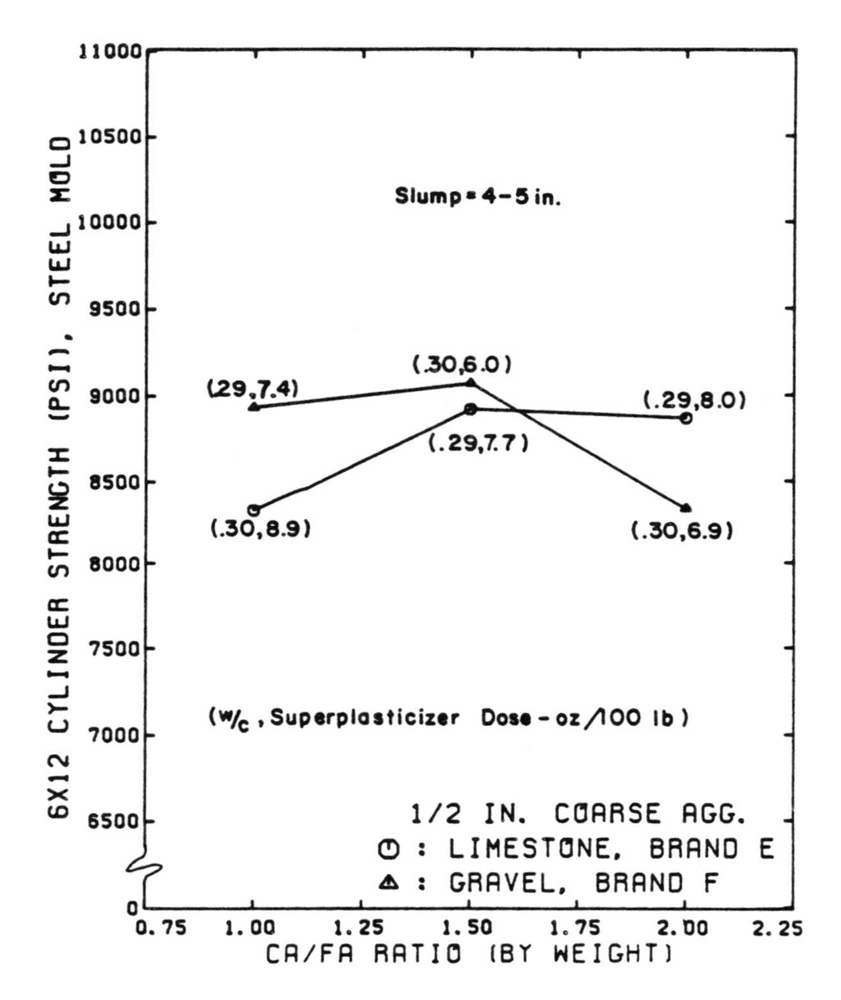

Fig. 4.49 Effect of coarse aggregate type and CA/FA ratio on the 28-day compressive strength of concrete for mixes having a cement content of 10 sacks/cu.yd. and made with type II cement, 1/2-in. max. size coarse aggregate, sand B, and superplasticizer B.

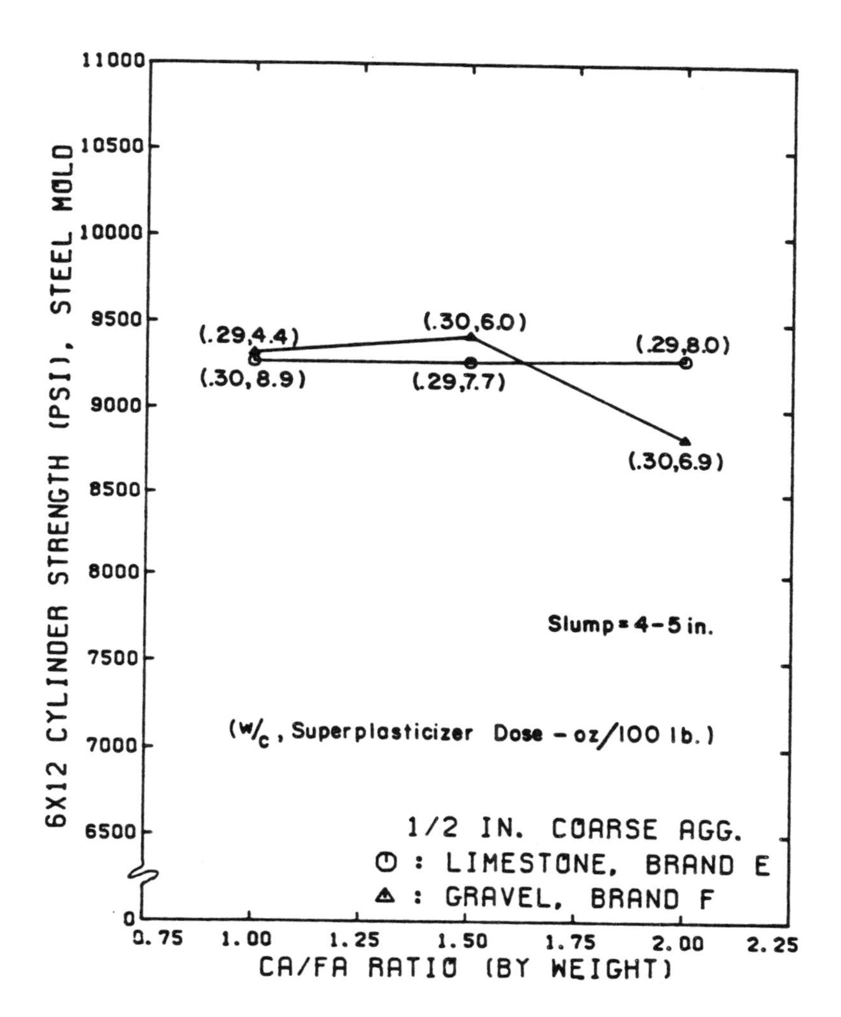

Fig. 4.50 Effect of coarse aggregate type and CA/FA ratio on the 56-day compressive strength of concrete for mixes having a cement content of 10 sacks/cu.yd. and made with type II cement, 1/2-in. max. size coarse aggregate, sand B, and superplasticizer B.

superplasticizers generally showed an increase in compressive strength at 28 and 56 days for increases in the CA/FA ratio. As shown in Figs. 4.48 and 4.50, high strength concrete was produced using any CA/FA ratio in the range from 1.0 to 2.0 using both gravel and limestone concrete.

4.8.3 Specimen Age. For mixes made with and without superplasticizers using all sizes and types of aggregate but containing no fly ash, the average compressive strength gain between 28 days and 56 days was 7.4%, as shown in Table 4.2. Crushed stone aggregates produced concrete with the highest rate of strength gain from 28 to 56 days, compared to gravel concretes. Concrete made with the limestone E had a lower rate of strength gain than did limestone D concrete.

In general, the rate of strength gain from 28 to 56 days was higher for mixes having a higher w/c ratio.

4.9 Sand Fineness

Three sands with fineness moduli ranging from 2.72 to 3.10 were used to compare the effects of sand fineness on concrete strength for high strength concrete containing no admixtures. Sands from the same source having fineness moduli ranging from 2.45 to 2.85 were used to compare mixes containing superplasticizers. In general, researchers have recommended the use of coarse sands for the production of high strength concrete. In addition, it is agreed that because of the high fines content of high strength concrete due to high cementitious content, the need for fine aggregate for finishability of fresh concrete is reduced.

4.9.1 Cement Content. The effect of sand fineness on strength of high strength concrete mixes made using 1/2-in. max. size coarse

TABLE 4.2 Comparison of the Average Rate of Increase in Compressive Strength of Concrete from a Test Age of 28 Days to 56 Days for Mixes made Using Different Types of Coarse Aggregate (includes No Mixes which Contain Fly Ash).

1/2-in. Coarse Aggregate	Gain in Compressive Strength of Concrete from 28 to 56 Days	
	With No Admixture	With Superplasticizer
Limestone E	13.0%	6.7%
Gravel F	10.8%	5.5%
Limestone D	18.5%	---

aggregate and different types of cement was studied. The concretes having the highest w/c ratios, for a given cement factor and CA/FA ratio had the lowest compressive strengths. However, the coarsest sands did not require the least mixing water for producing high-strength concrete with a 3-in. to 4-in. slump. As shown in Figs. 4.51 through 4.54, the mixes made using the finest sand had the highest 28-day compressive strength for a given CA/FA ratio. A similar relationship was observed for mixes made with cement types I and III and other CA/FA ratios.

Fineness modulus had little effect on 56-day compressive strength for any CA/FA ratio in mixes containing 10 sacks/cu.yd., but the finest aggregate produced the strongest concrete in 7-sack mixes. The w/c ratio controlled strength at 28 days but seems to have had little direct relationshp with 56-day strengths, as seen in Figs. 4.52, 4.53, and 4.54.

Mixes containing superplasticizer seemed to be controlled by admix dosage as much as by sand fineness.

The effects of sand fineness on compressive strength of concrete mixes containing superplasticizer can be seen in Figs. 4.55 through 4.57. Finer sands generally produced higher strength concrete for mixes containing 8.5 sacks of cement/cu.yd. and any type of coarse aggregate at any CA/FA ratio. However, these mixes made with finer sands also required the greatest dose of superplasticizer, so the resulting strength increase could have been affected by the increased superplasticizer dosages.

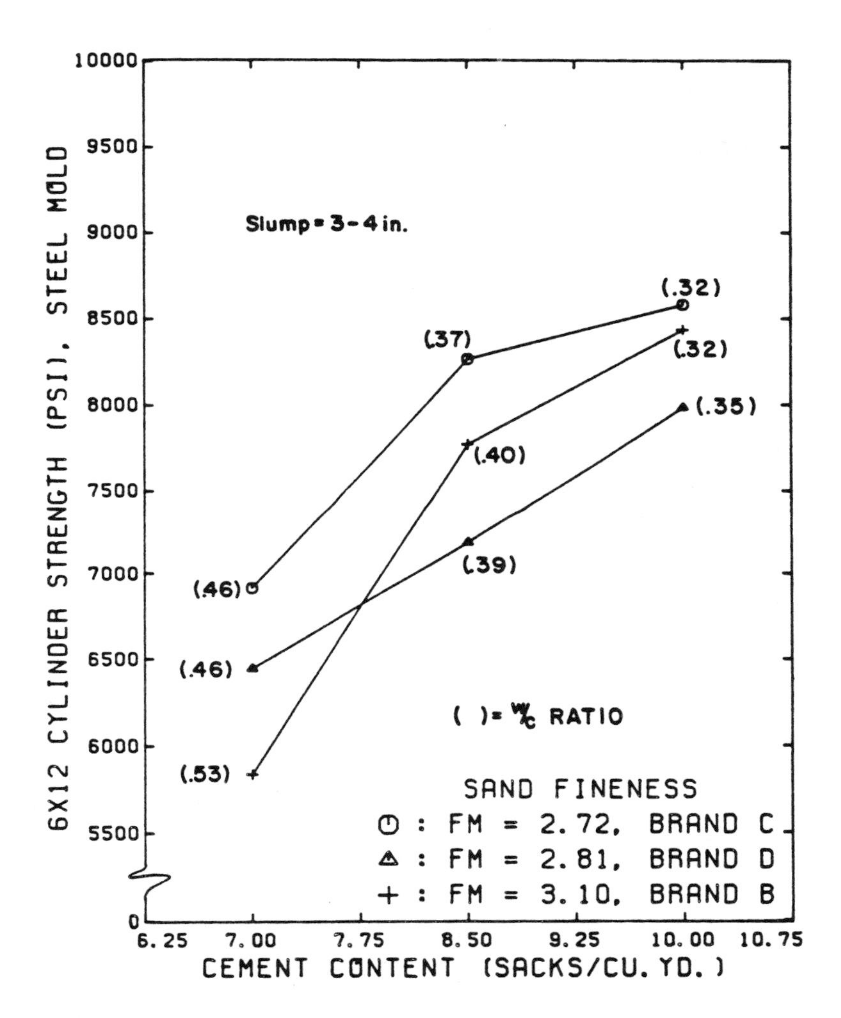

Fig. 4.51 Effect of sand fineness and cement content on the 28-day compressive strength of concrete for mixes having a CA/FA ratio of 2.0 and made with type II cement, 1/2-in. limestone E, and no admixture.

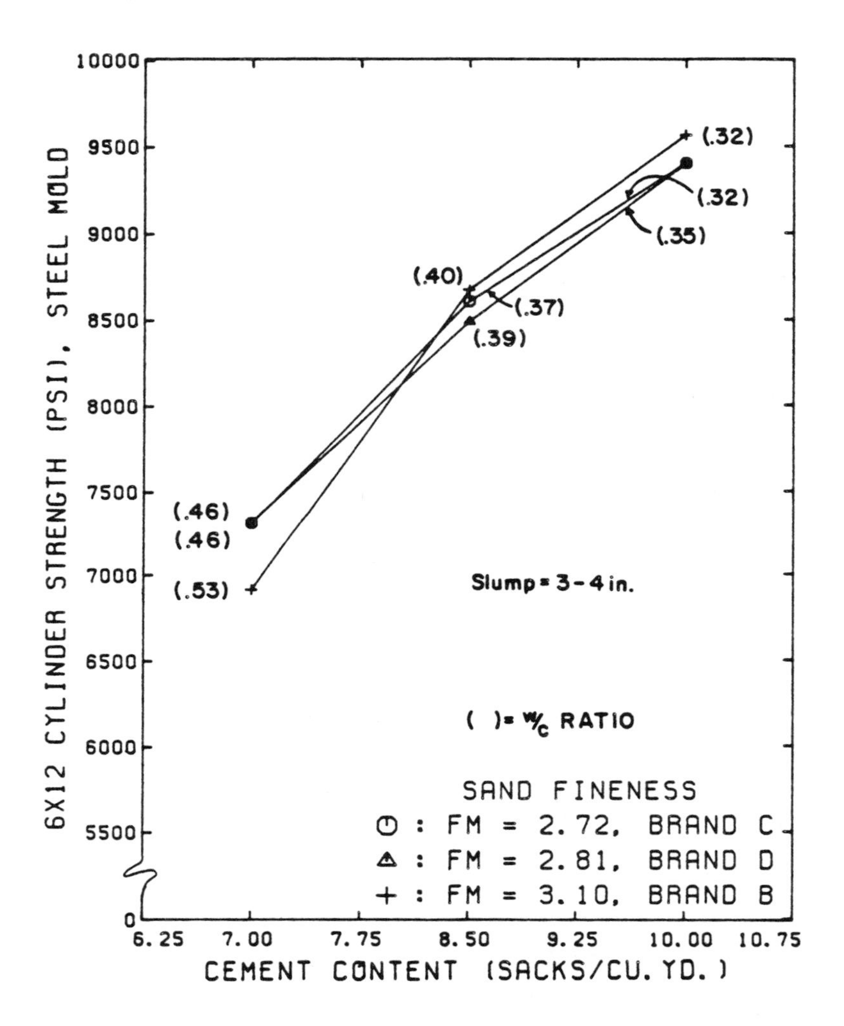

Fig. 4.52 Effect of sand fineness and cement content on the 56-day compressive strength of concrete for mixes having a CA/FA ratio of 2.0 and made with type II cement, 1/2-in. limestone E, and no admixture.

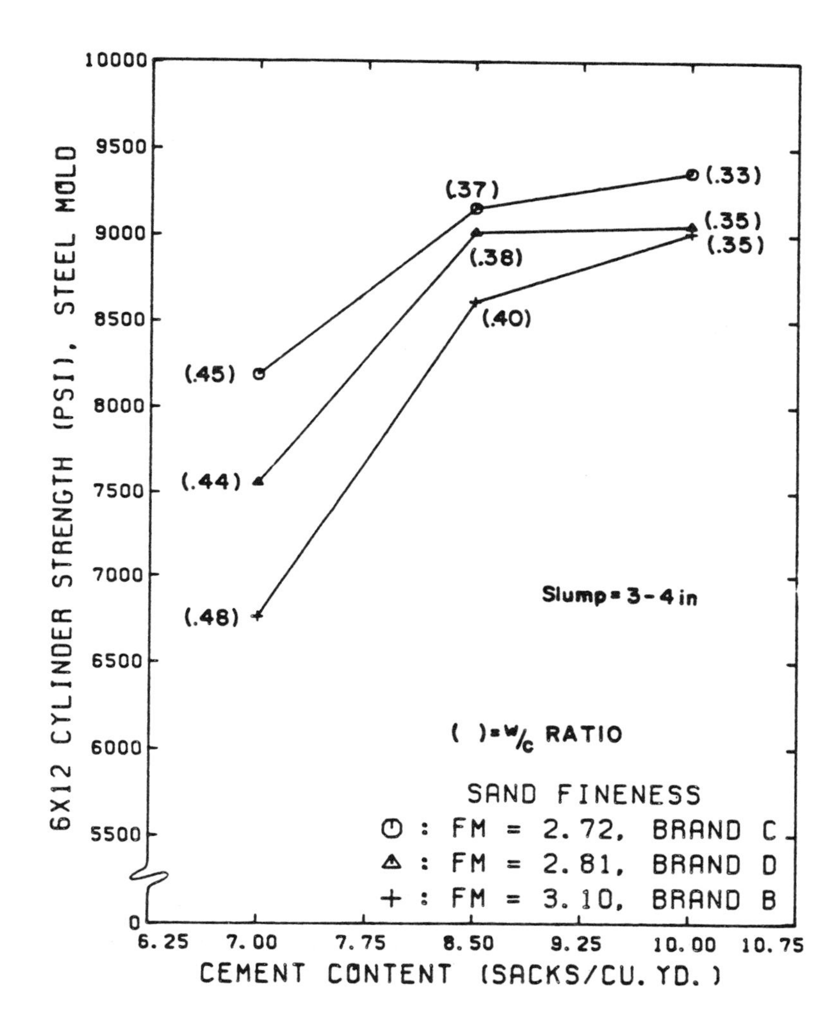

Fig. 4.53 Effect of sand fineness and cement content on the 56-day compressive strength of concrete for mixes having a CA/FA ratio of 1.5 and made with type II cement, 1/2-in. limestone E, and no admixture.

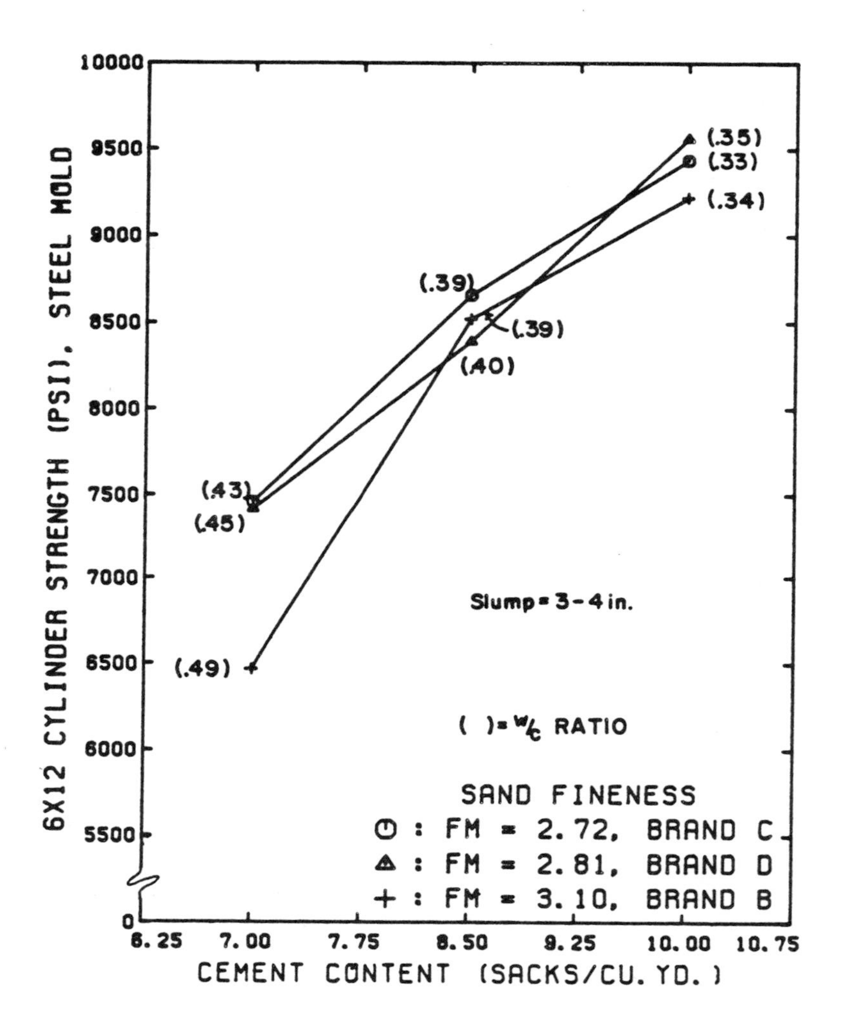

Fig. 4.54 Effect of sand fineness and cement content on the 56-day compressive strength of concrete for mixes having a CA/FA ratio of 1.0 and made with type II cement, 1/2-in. limestone E, and no admixture.

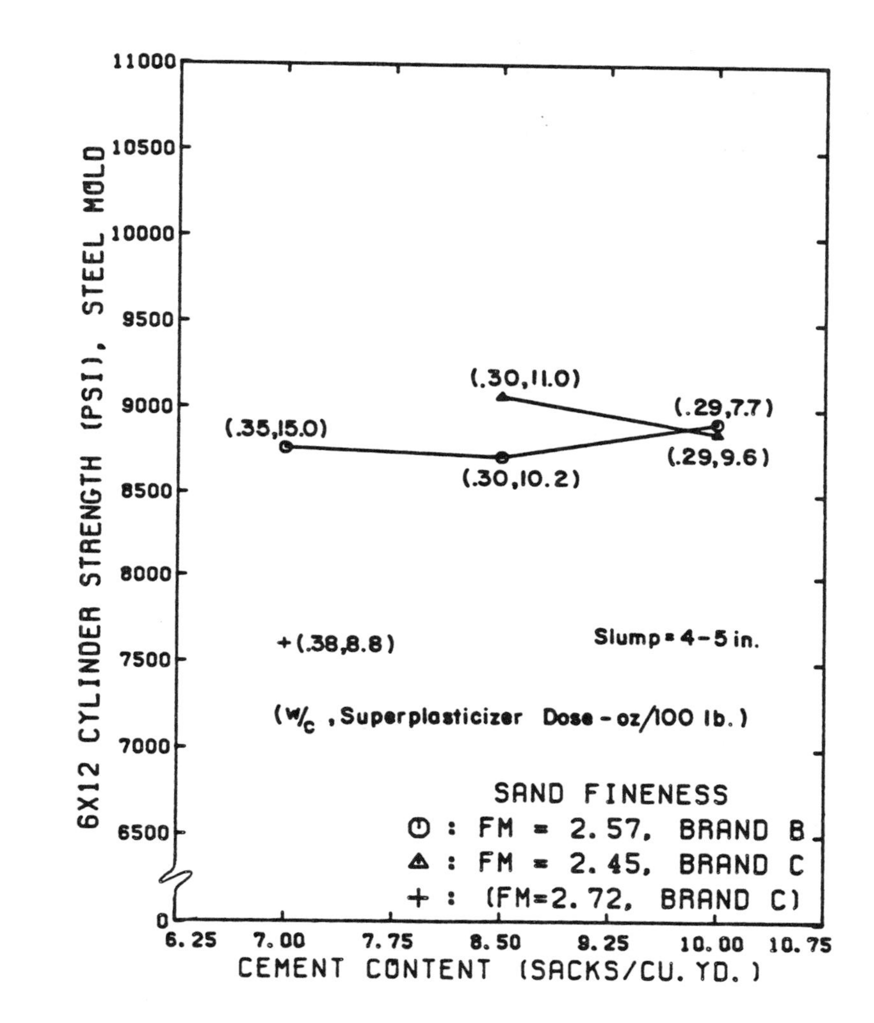

Fig. 4.55 Effect of sand fineness and cement content on the 28-day compressive strength of concrete for mixes having a CA/FA ratio of 1.5 and made with type II cement, 1/2-in. limestone E, and superplasticizer B.

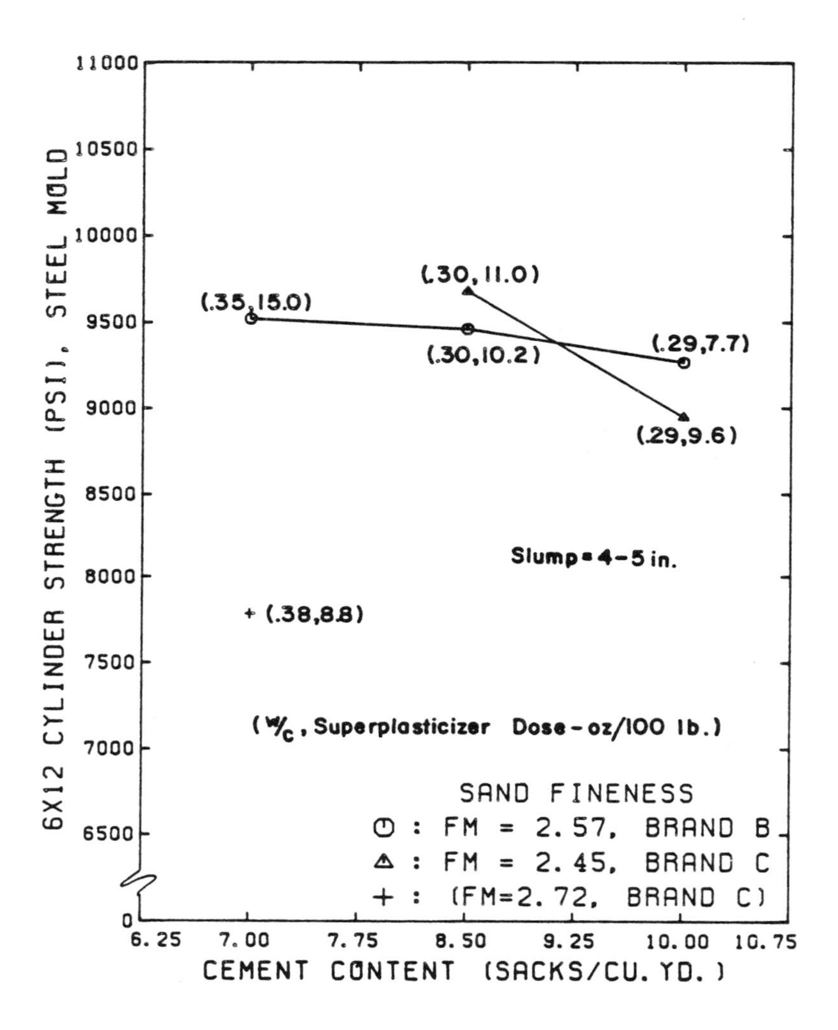

Fig. 4.56 Effect of sand fineness and cement content on the 56-day compressive strength of concrete for mixes having a CA/FA ratio of 1.5 and made with type II cement, 1/2-in. limestone E, and superplasticizer B.

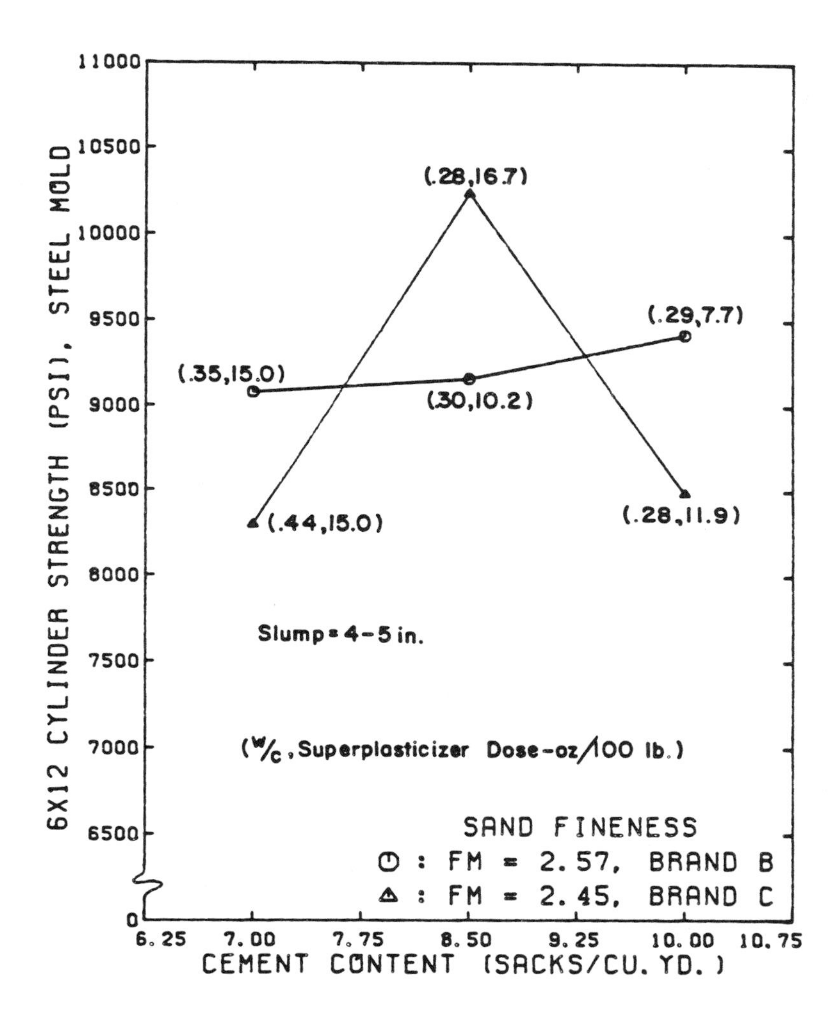

Fig. 4.57 Effect of sand fineness and cement content on the 56-day compressive strength of concrete for mixes having a CA/FA ratio of 1.5 and made with type II cement, 1/2-in. gravel F, and superplasticizer B.

Compressive strength of concrete mixes containing 10 sacks of cement/cu.yd. was adversely affected by use of finer sands. The excessive fines content tended to reduce compressive strength regardless of superplasticizer dosage. For a cement content of 7.0 sacks/cu.yd. higher compressive strengths were obtained in mixes made using the coarser sand. The w/c ratio in these mixes were lower for a given superplasticizer dosage.

In general, high strength concrete can be produced using sands having a fineness modulus as low as 2.45 if superplasticizer is used.

4.9.2 Coarse/Fine Aggregate Ratio. As shown in Figs. 4.58 through 4.60, there was no clear trend between compressive strength of concrete as a function of the CA/FA ratio and sand fineness in mixes containing no admixtures. It can be seen by comparisons between these figures that the difference in compressive strength due to a change in sand fineness was reduced as cement content increased regardless of the CA/FA ratio.

Figure 4.61 shows a typical plot of compressive strength versus CA/FA ratio for mixes containing superplasticizer and different sands. In general, a CA/FA ratio of 2.0 produced the highest compressive strengths regardless of sand fineness.

4.9.3 Specimen Age. For mixes made with cement contents of 7 to 8.5 sacks/cu.yd., but containing no admixtures, using the coarsest sand resulted in an increase in w/c ratio and the lowest compressive strength at any age for all CA/FA ratios, as shown by Figs. 4.62 and 4.63. As seen in Fig. 4.63, a difference of 500 to 1,500 psi in

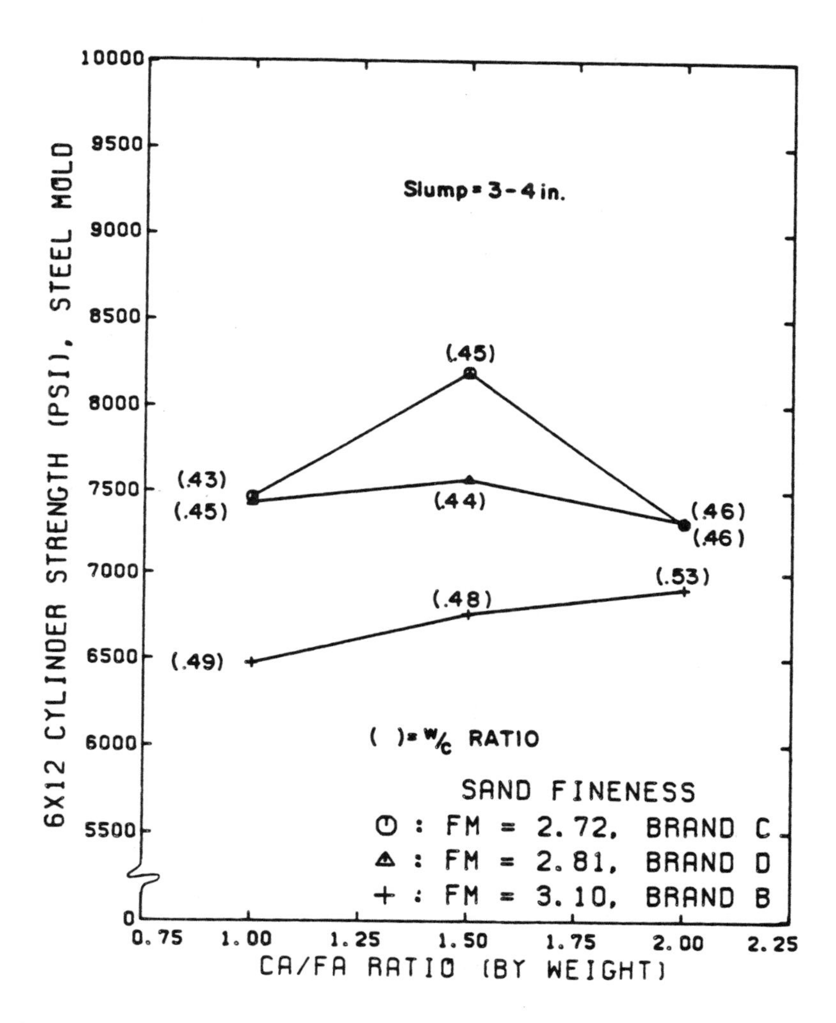

Fig. 4.58 Effect of sand fineness and CA/FA ratio on the 56-day compressive strength of concrete for mixes having a cement content of 7.0 sacks/cu.yd. and made with type II cement, 1/2-in. limestone E, and no admixture.

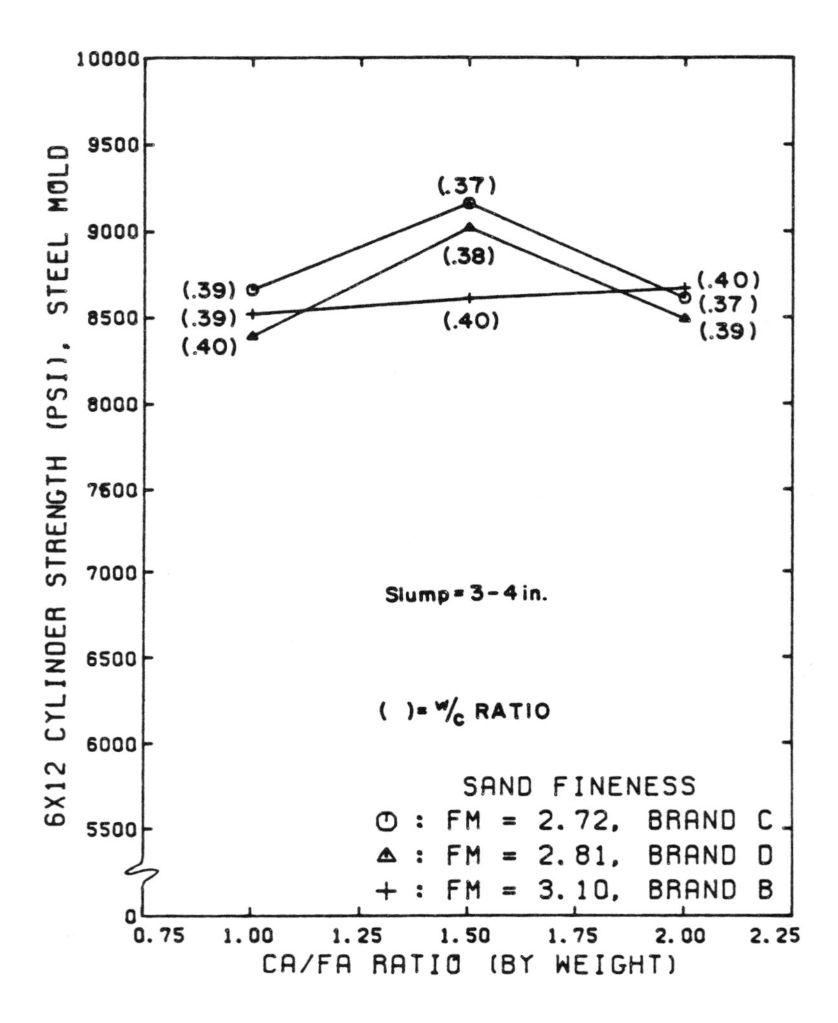

Fig. 4.59 Effect of sand fineness and CA/FA ratio on the 56-day compressive strength of concrete for mixes having a cement content of 8.5 sacks/cu.yd. and made with type II cement, 1/2-in. limestone E, and no admixture.

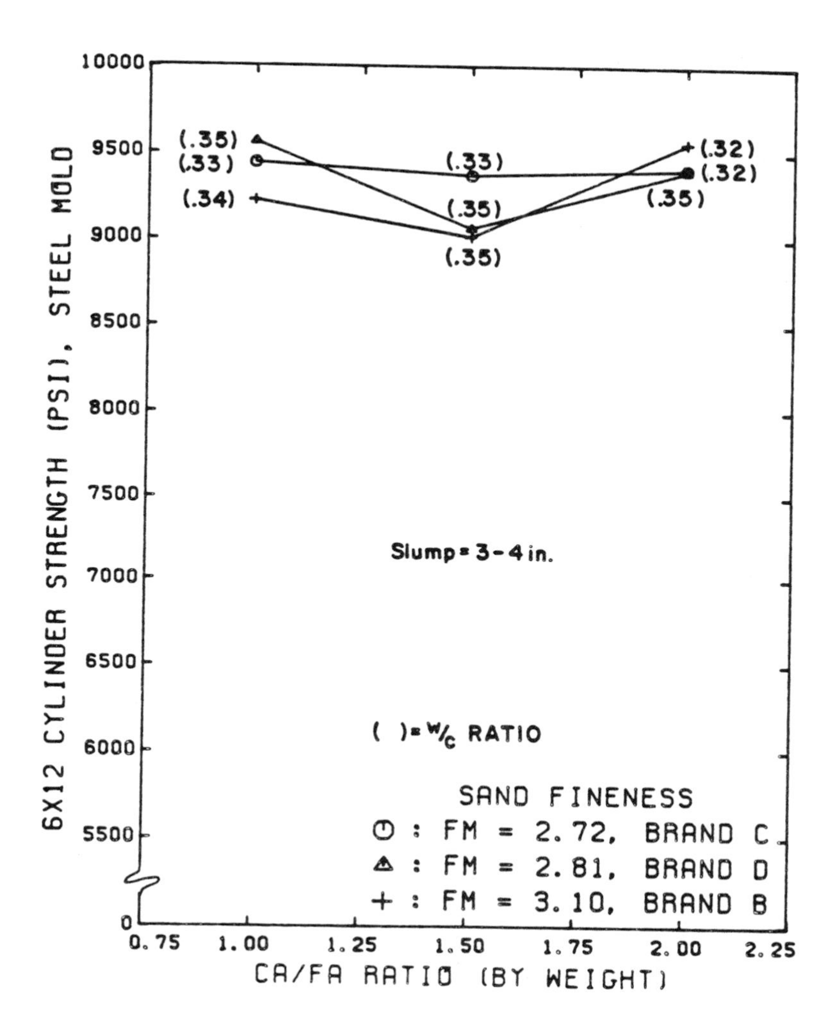

Fig. 4.60 Effect of sand fineness and CA/FA ratio on the 56-day compressive strength of concrete for mixes having a cement content of 10 sacks/cu.yd. and made with type II cement, 1/2-in. limestone E, and no admixture.

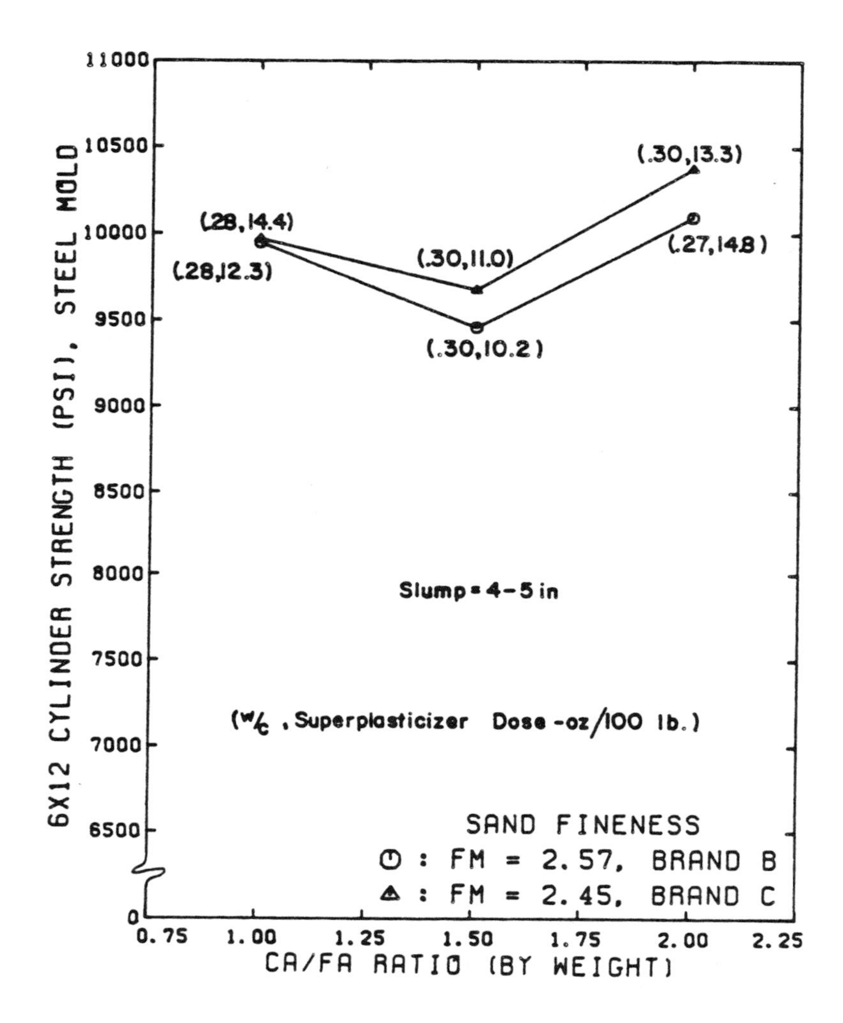

Fig. 4.61 Effect of sand fineness and CA/FA ratio on the 56-day compressive strength of concrete for mixes having a cement content of 8.5 sacks/cu.yd. and made with type II cement, 1/2-in. limestone E, and superplasticizer B.

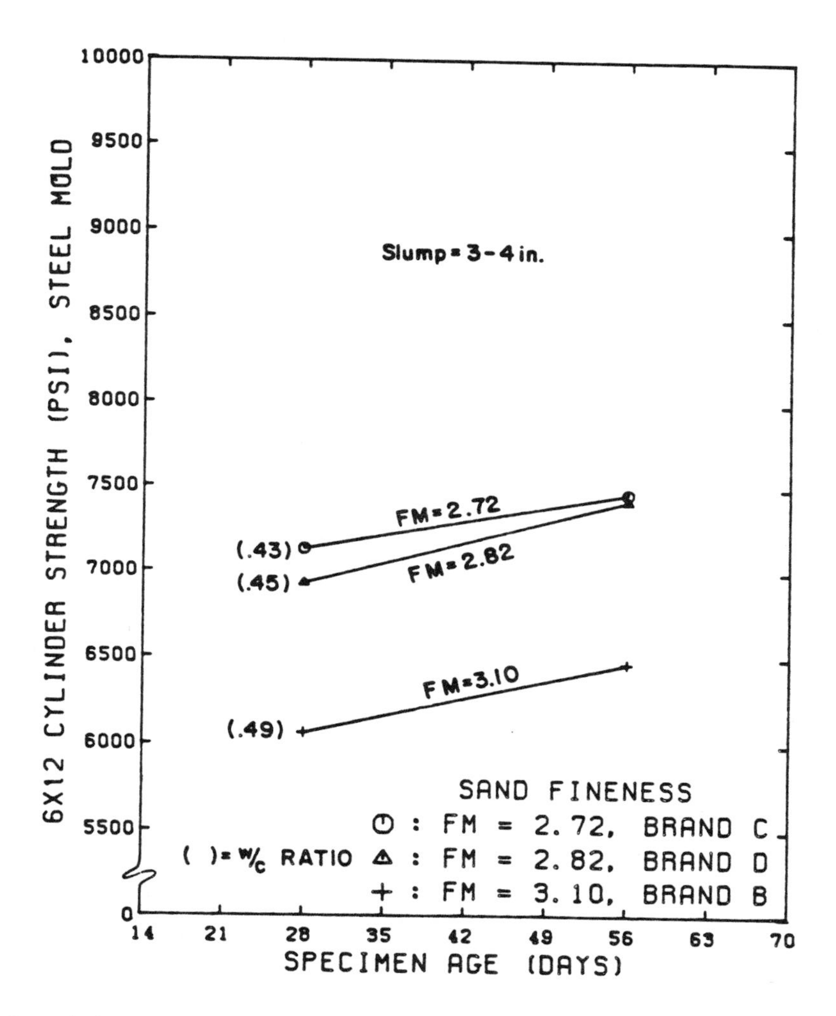

Fig. 4.62 Effect of sand fineness and specimen age on the compressive strength of concrete for mixes having a cement content of 7.0 sacks/cu.yd. and a CA/FA ratio of 1.0 and made with type II cement, 1/2-in. limestone E, and no admixture.

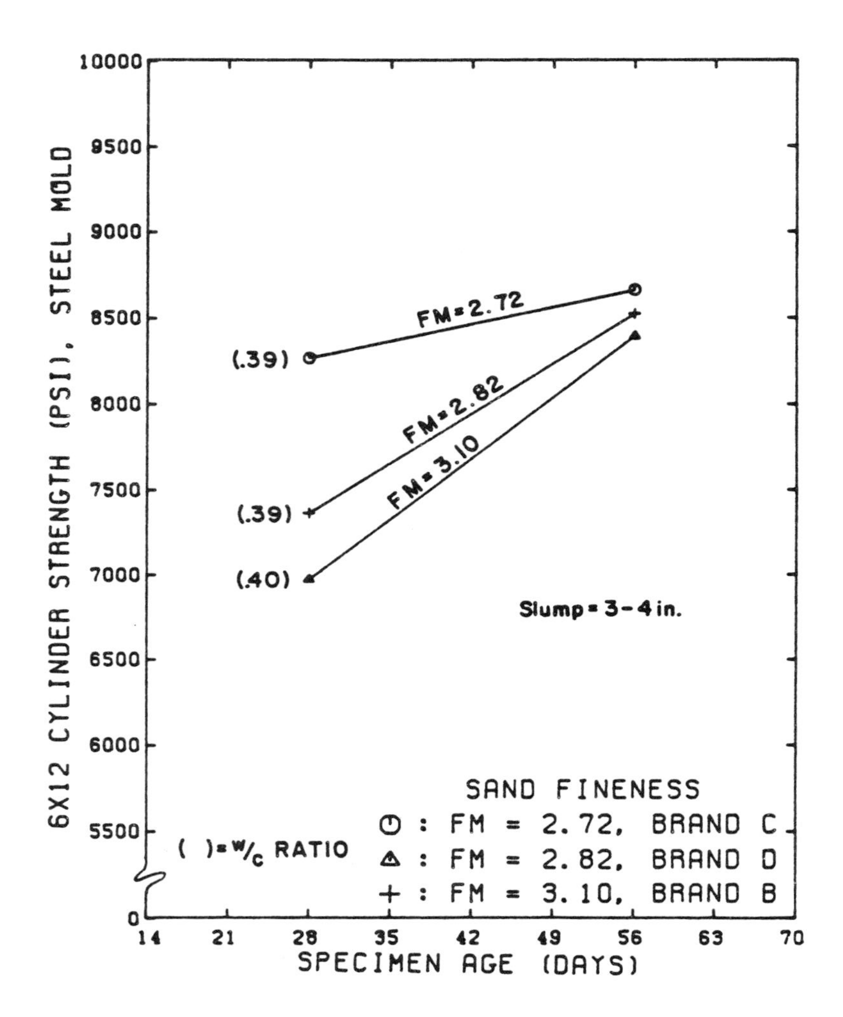

Fig. 4.63 Effect of sand fineness and specimen age on the compressive strength of concrete for mixes having a cement content of 8.5 sacks/cu.yd. and a CA/FA ratio of 1.0 and made with type II cement, 1/2-in. limestone E, and no admixture.

compressive strength existed between the strongest concrete and the weakest at any age. Figures 4.63 through 4.66 show that for mixes having cement contents of at least 8.5 sacks/cu.yd., higher w/c ratios produced lower 28-day compressive concrete strength but resulted in a higher rate of strength gain up to a test age of 56 days.

It can be seen that sand fineness had no consistent effect on 56-day compressive strengths for different CA/FA ratios in mixes containing 10 sacks/cu.yd. However, the finest aggregate consistently produced the strongest concrete at 28 days.

Figures 4.67 and 4.68 show that in concrete containing superplasticizers finer sands resulted in higher compressive strengths in 8.5-sack mixes and lower compressive strengths in 10-sack mixes at any age. The rate of strength gain with curing age from 28 to 56 days was higher for lower superplasticizer dosages.

4.10 Fly Ash

The addition of fly ash to high strength concrete mixes increased the compressive strength at 28 and 56 days more than did the addition of the same weight of Portland cement. Substituting class C fly ash for 20 to 30 percent of the cement in a mix containing no chemical admixtures resulted in concretes having 28-day compressive strengths of nearly 10,000 psi.

In this report, "percent fly ash" refers to the ratio by weight of fly ash to total binder (Portland cement plus fly ash) expressed as a percent. The term "w/b" is the ratio by weight of total required mixing

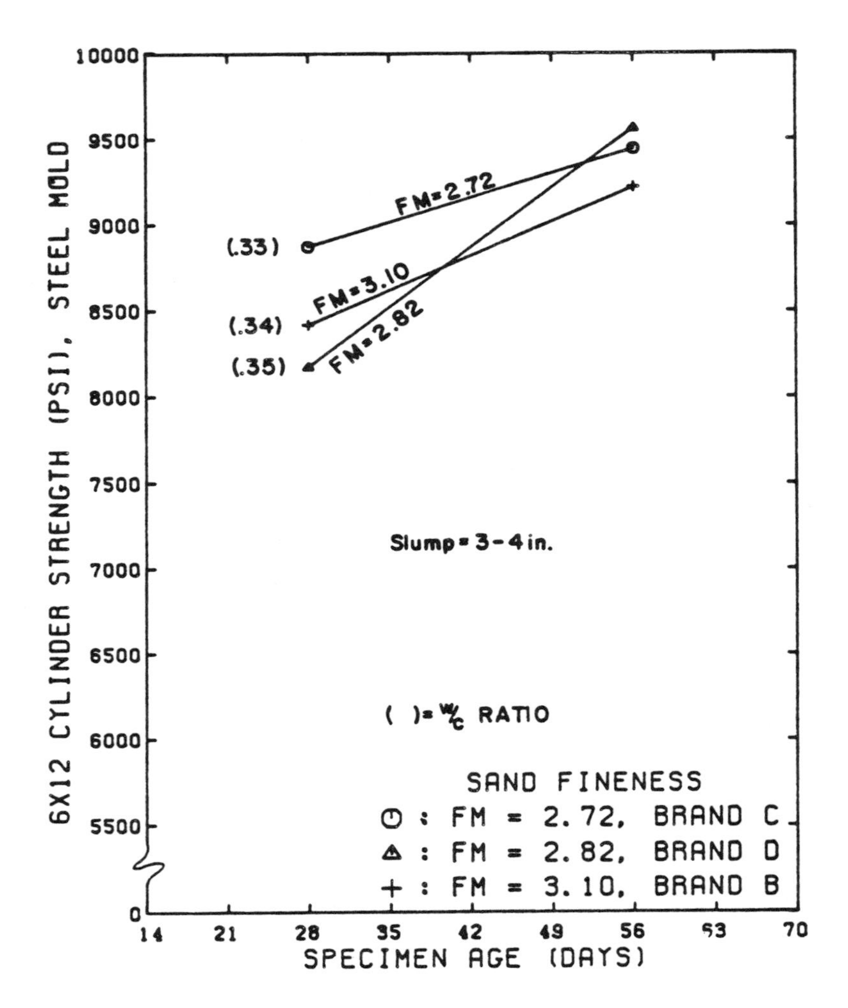

Fig. 4.64 Effect of sand fineness and specimen age on the compressive strength of concrete for mixes having a cement content of 10 sacks/cu.yd. and a CA/FA ratio of 1.0 and made with type II cement, 1/2-in. limestone E, and no admixture.

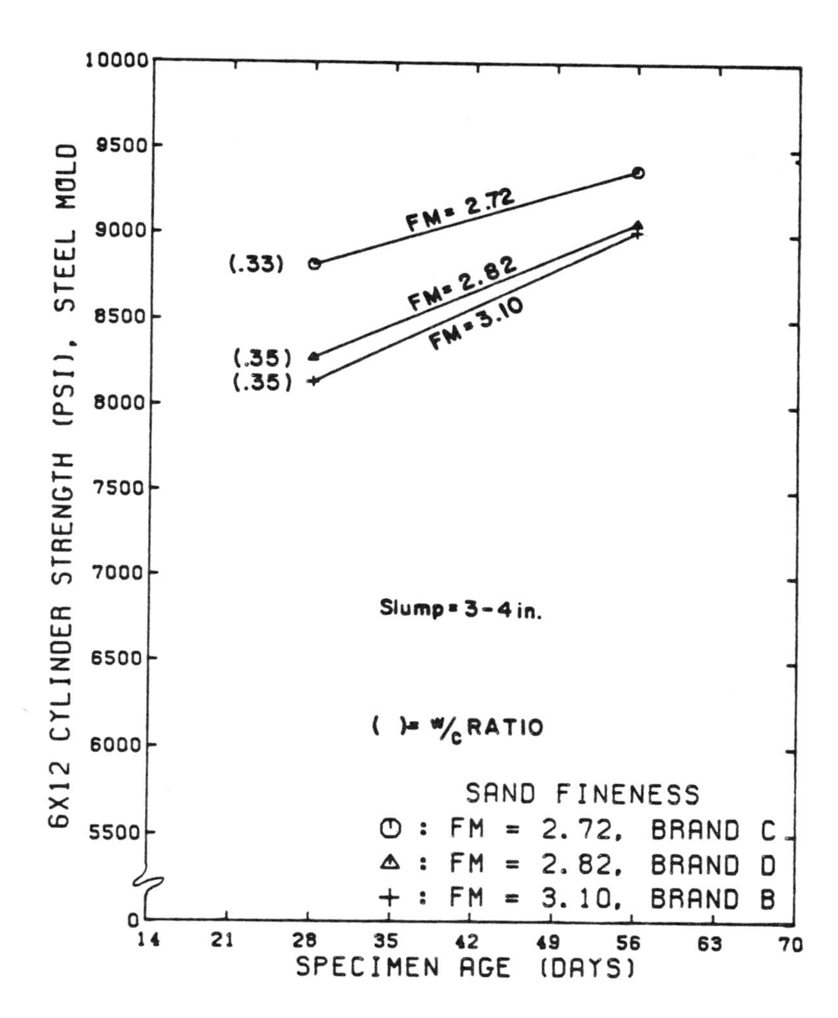

Fig. 4.65 Effect of sand fineness and specimen age on the compressive strength of concrete for mixes having a cement content of 10 sacks/cu.yd. and a CA/FA ratio of 1.5 and made with type II cement, 1/2-in. limestone E, and no admixture.

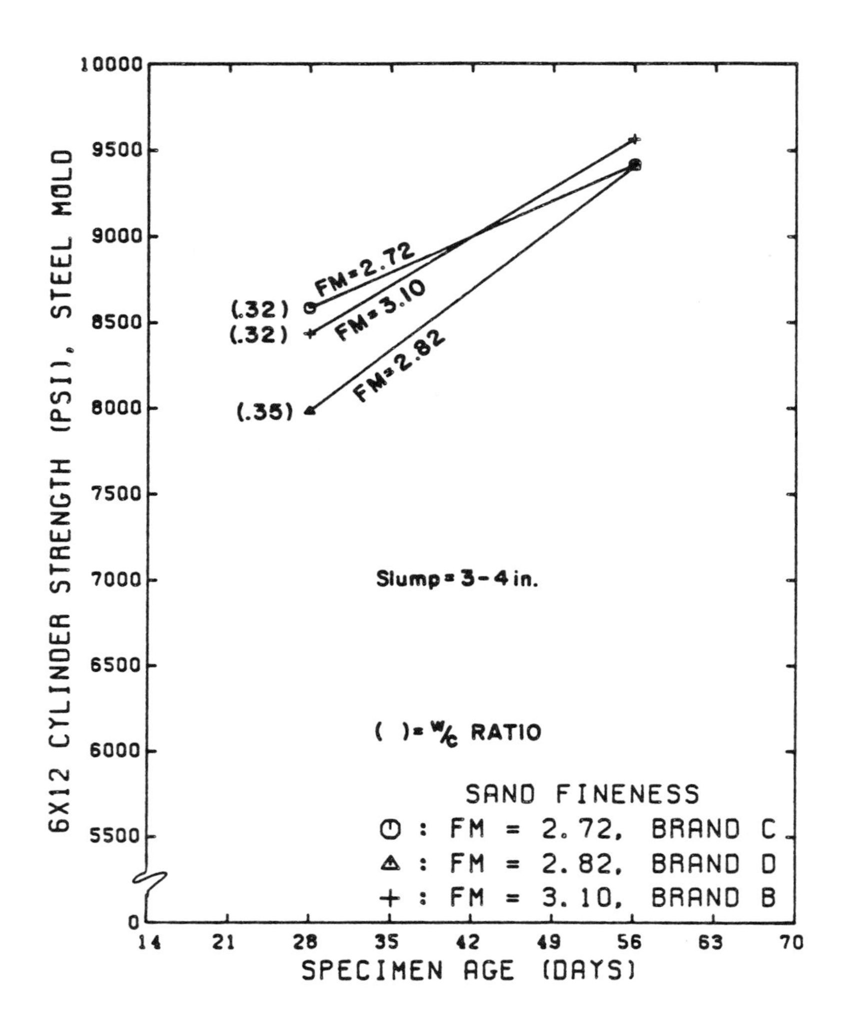

Fig. 4.66 Effect of sand fineness and specimen age on the compressive strength of concrete for mixes having a cement content of 10 sacks/cu.yd. and a CA/FA ratio of 2.0 and made with type II cement, 1/2-in. limestone E, and no admixture.

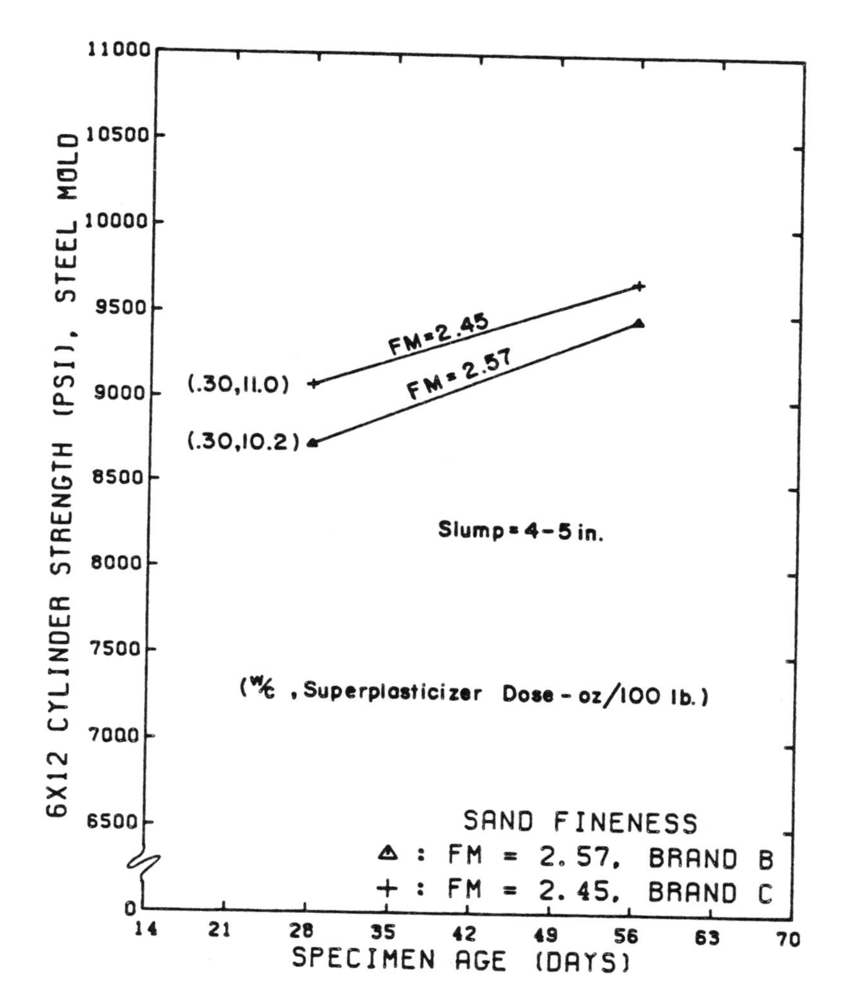

Fig. 4.67 Effect of sand fineness and specimen age on the compressive strength of concrete for mixes having a cement content of 8.5 sacks/cu.yd. and a CA/FA ratio of 1.5 and made with type II cement, 1/2-in. limestone E, and superplasticizer B.

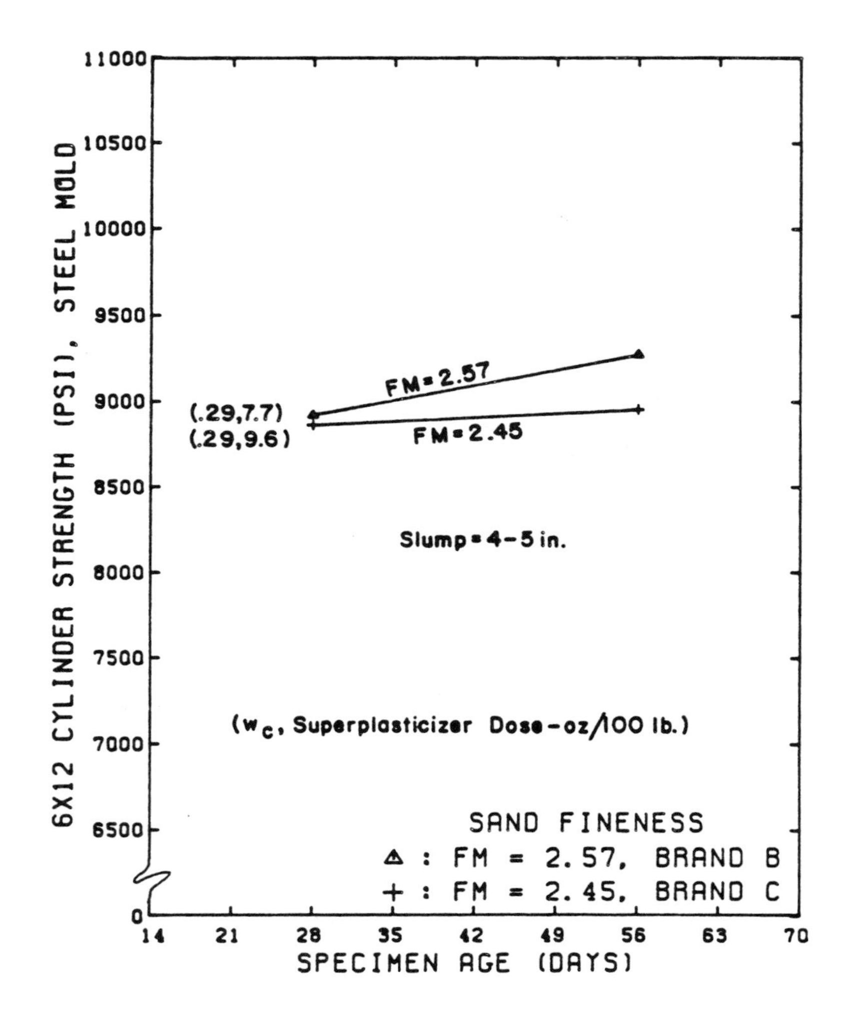

Fig. 4.68 Effect of sand fineness and specimen age on the compressive strength of concrete for mixes having a cement content of 10 sacks/cu.yd. and a CA/FA ratio of 1.5 and made with type II cement, 1/2-in. limestone E, and superplasticizer B.

water to total binder. Chemical admixture dosages are reported as fluid ounces of admixture per 100 lbs of Portland cement.

4.10.1 Total Cementitious Materials Content. Compressive strengths at 28 and 56 days are plotted in Figs. 4.69 and 4.70 against total cementitious material content, or total binder weight, for different fly ash contents. At 28 days, the concrete mixes with the higher fly ash content resulted in higher compressive strength. As shown in Fig. 4.69, mixes containing 30 percent fly ash had the highest compressive strength at 28 days. For mixes containing no chemical admixtures, the highest compressive strength was achieved by using approximately 1,000 lbs of binder per cubic yard for fly ash contents ranging from 20 percent to 30 percent.

However, for a total binder weight of more than 1,000 lbs per cu.yd., there was little difference in compressive strength between mixes having a fly ash to total binder ratio of 20 percent to 30 percent by weight.

At 56 days, mixes containing 0 percent to 20 percent fly ash showed a signficant strength increase over the 28 day strength. Mixes containing 30 percent fly ash showed little or no strength increase for the same test age except for mixes having a total binder content of less than 950 lbs per cu.yd. As a result, mixes havng a ratio of fly ash to total binder of 20 percent produced the highest concrete compressive strengths at 56 days. The 56 day strengths were more closely related to the water/binder ratio. At 56 days, for a given total binder content, the mixes with lowest w/b ratio also produced the highest compressive

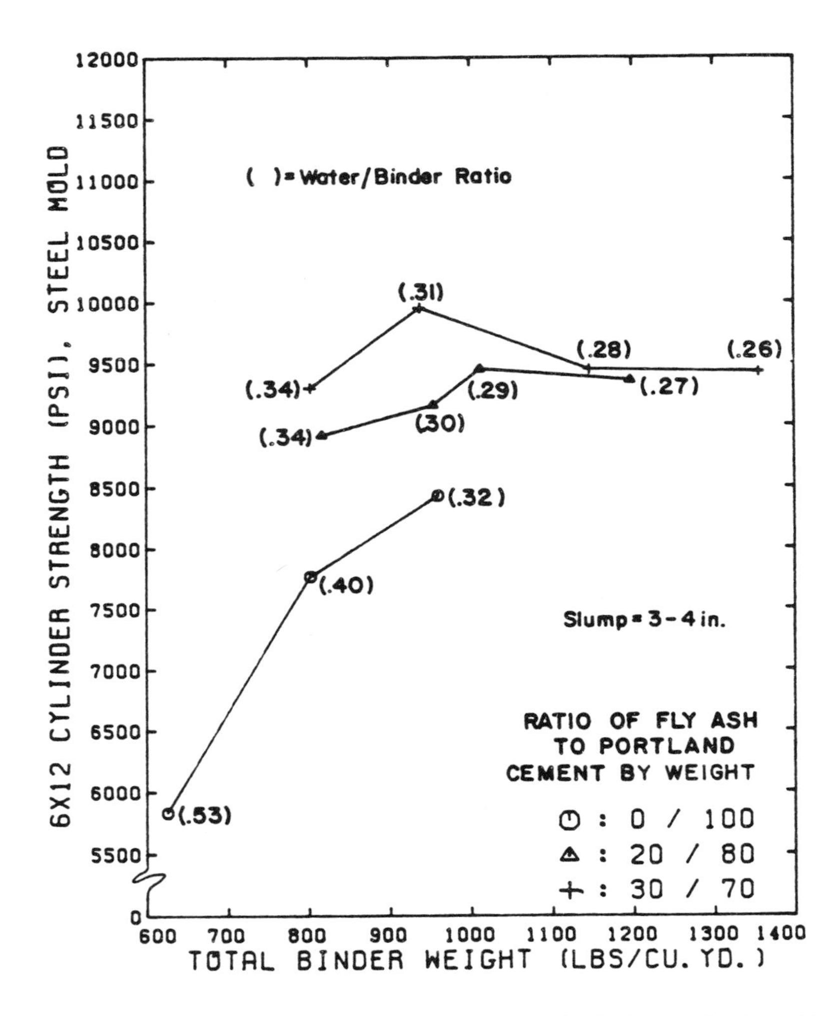

Fig. 4.69 Effect of fly ash content and total binder content on the 28-day compressive strength of concrete for mixes having a CA/FA ratio of 2.0 and made with type II cement, fly ash A, 1/2-in. limestone E, sand B, and no chemical admixture.

the mixes with lowest w/b ratio also produced the highest compressive strengths, as seen in Fig. 4.70.

In any case, Figs. 4.69 and 4.70 clearly show that for a given total binder weight per cu.yd., the concrete compressive strength increases if 20 to 30 percent of the total weight of Portland cement was replaced by an equal weight of Class C fly ash.

Compressive strength of concrete at 28 and 56 days is plotted versus fly ash content expressed as a percentage of the total weight of binder in Figs. 4.71 and 4.72.

As shown in these figures, increasing the total binder content tended to increase the 28 day compressive strength of concrete for any fly ash content up to at least 30 percent. However, for mixes containing 30 percent fly ash, an increase in total binder content beyond 800 lbs per cu.yd. did not result in any increase in strength after 28 days.

Adding fly ash to mixes containing superplasticizer did not result in as great an increase in compressive strength as did the addition of fly ash to mixes containing no chemical admixtures. Strength gains of approximately 10 percent resulted from the addition of fly ash to mixes containing superplasticizers. As had been the case in earlier comparisons of mixes with and without superplasticizers, trends of compressive strength as a function of binder content were not well defined, as shown in Fig. 4.73. The highest compressive strength was obtained with a concrete mix containing 800 lbs of binder with a 30 percent fly ash content.

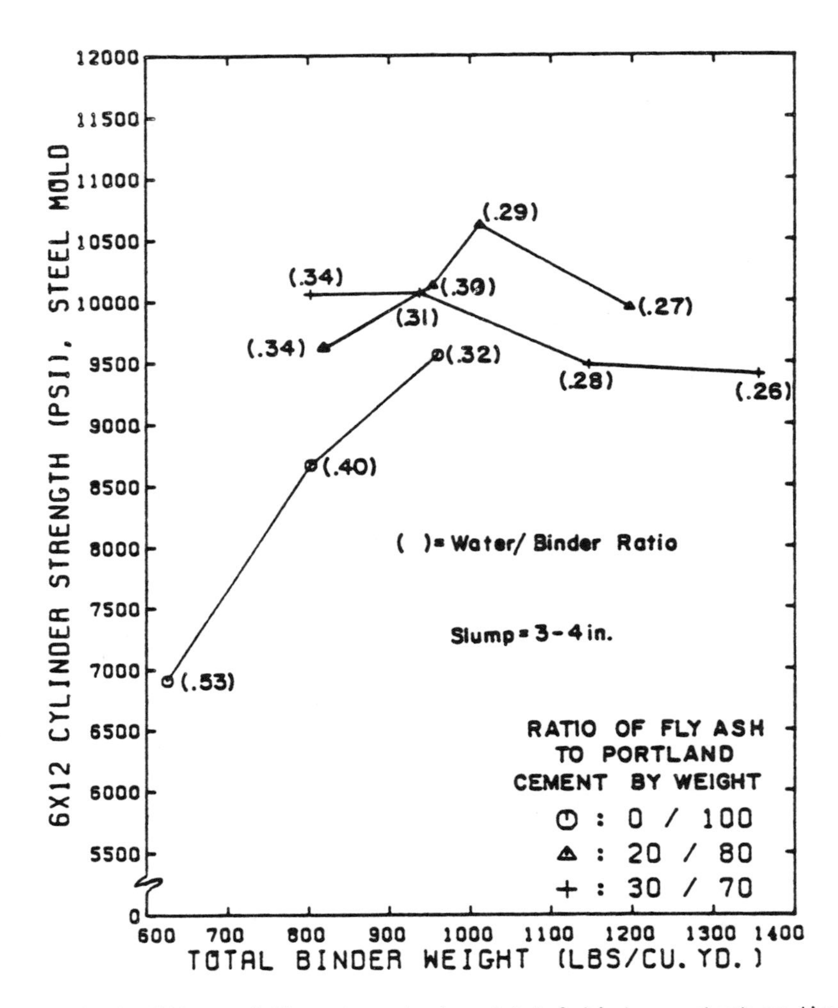

Fig. 4.70 Effect of fly ash content and total binder content on the 56-day compressive strength of concrete for mixes having a CA/FA ratio of 2.0 and made with type II cement, fly ash A, 1/2-in. limestone E, sand B, and no chemical admixture.

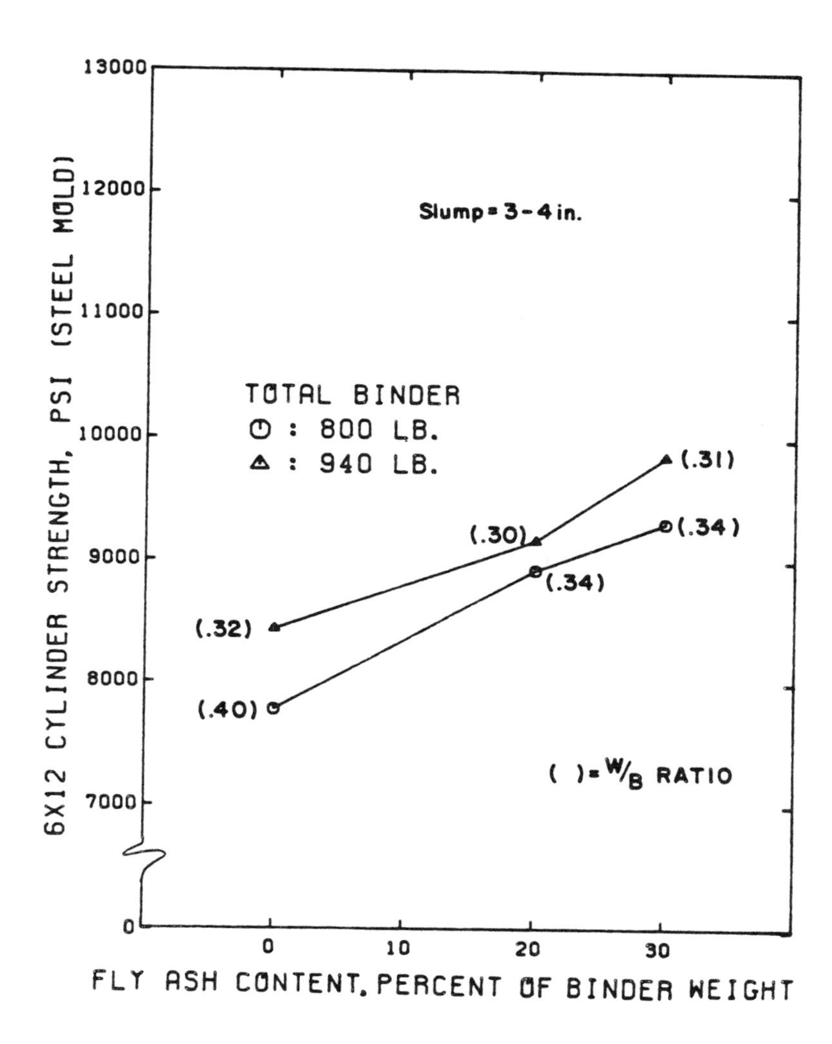

Fig. 4.71 Effect of total binder content and fly ash content on the 28-day compressive strength of concrete for mixes having a CA/FA ratio of 2.0 and made with type II cement, fly ash A, 1/2-in. limestone E, sand B, and no admixture.

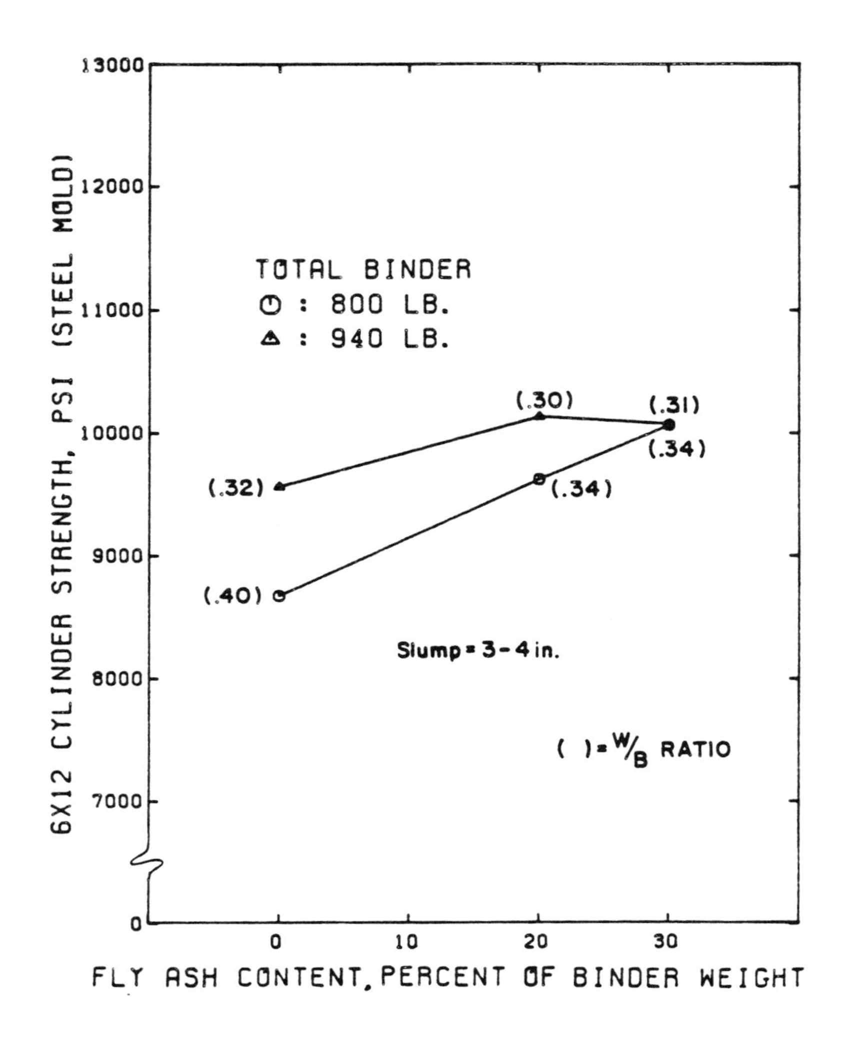

Fig. 4.72 Effect of total binder content and fly ash content on the 56-day compressive strength of concrete for mixes having a CA/FA ratio of 2.0 and made with type II cement, fly ash A, 1/2-in. limestone E, sand B, and no admixture.

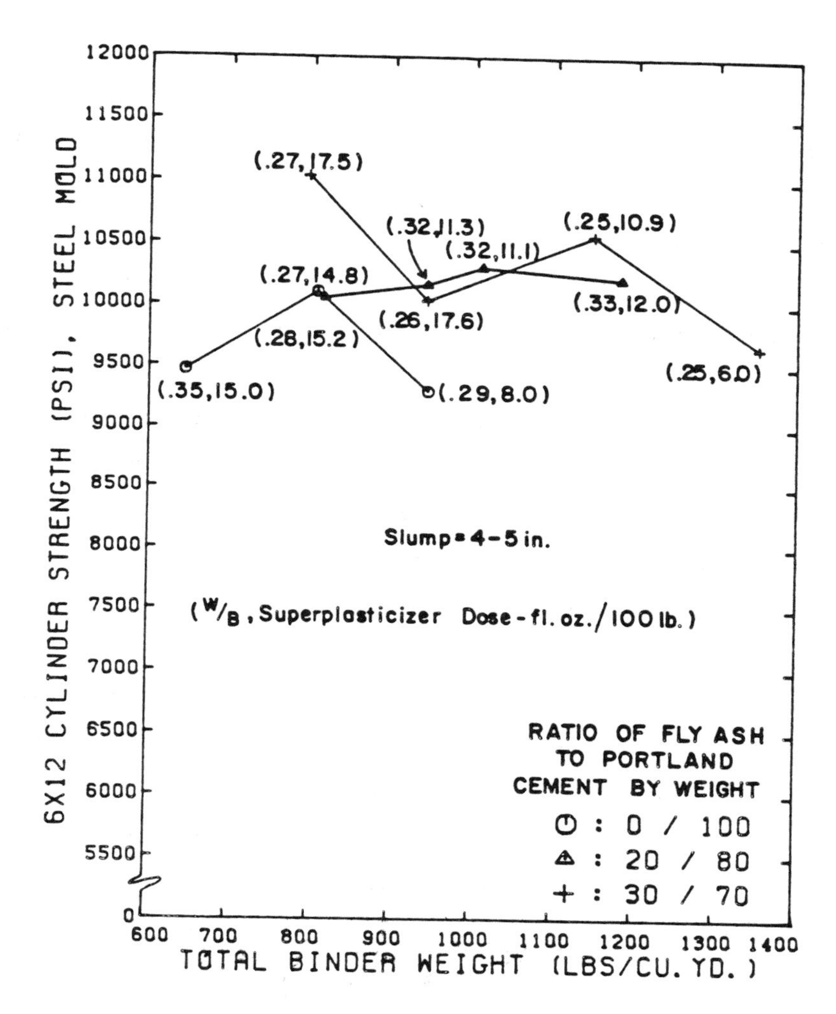

Fig. 4.73 Effect of fly ash content and total binder content on the 56-day compressive strength of concrete for mixes having a CA/FA ratio of 2.0 and made with type II cement, fly ash A, 1/2-in. limestone E, sand B, and superplasticizer B.

A direct comparison of compressive strength for identical concrete mixes containing 30 percent fly ash, with and without superplasticizers, is shown in Figs. 4.74 and 4.75. In general, the addition of superplasticizers did not significantly affect compressive strength at 28 days except for the leanest mix which required a higher admixture dosage for a given slump. At 56 days, mixes containing superplasticizer generally resulted in higher compressive strengths than mixes containing no chemical admixture.

In Fig. 4.76 it is shown that little or no compressive strength was gained at 28 days by using more than 800 lbs of binder per cu.yd. (8.5 "sacks"/cu.yd.) for mixes containing superplasticizer. Mixes having the highest superplasticizer dosages resulted in the highest compressive strengths for a given fly ash content.

4.10.2 Fly Ash Source. The effects of using a class C fly ash from two different sources was also studied. Mixes with and without superplasticizers were made with fly ash contents of 20 and 30 percent. Compressive strength test results from one set of mixes are shown in Figs. 4.77 and 4.78. For mixes containing superplasticizer with 6 sacks of Portland cement plus 240 lbs of fly ash, the compressive strength and rate of strength increase were affected by fly ash brand. In this case, fly ash B produced lower strength at 28 days but resulted in the highest 56-day compressive strength. However, the workability and compressive strength of concrete made with fly ash B were similar to those made using fly ash A.

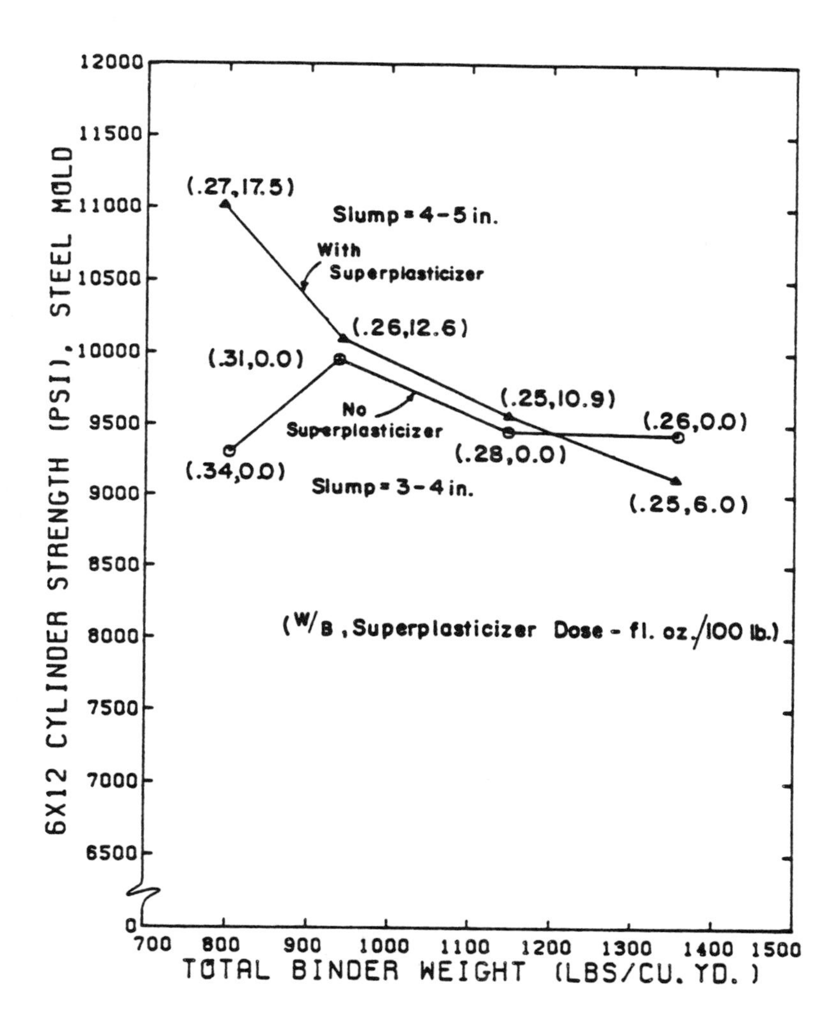

Fig. 4.74 Effect of superplasticizer and total binder content on the 28-day compressive strength of concrete for mixes having a CA/FA ratio of 2.0 and made with type II cement, fly ash A (30% by wt.), 1/2-in. limestone E, and sand B.

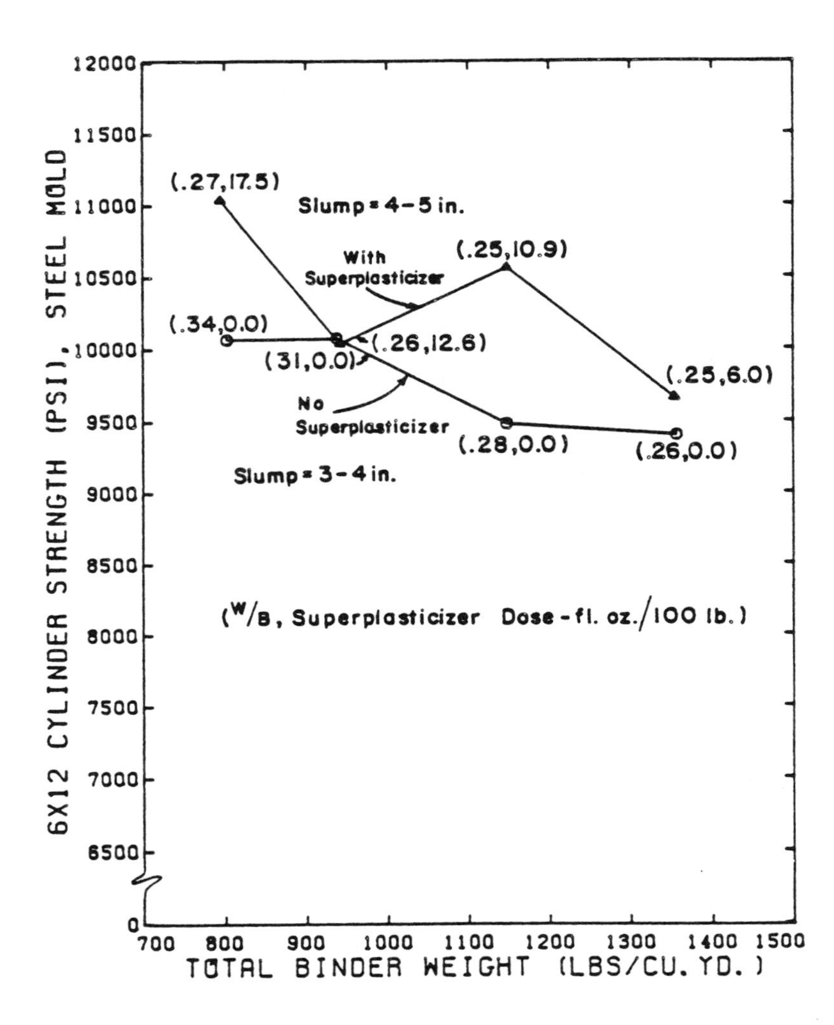

Fig. 4.75 Effect of superplasticizer and total binder content on the 56-day compressive strength of concrete for mixes having a CA/FA ratio of 2.0 and made with type II cement, fly ash A (30% by wt.), 1/2-in. limestone E, and sand B.

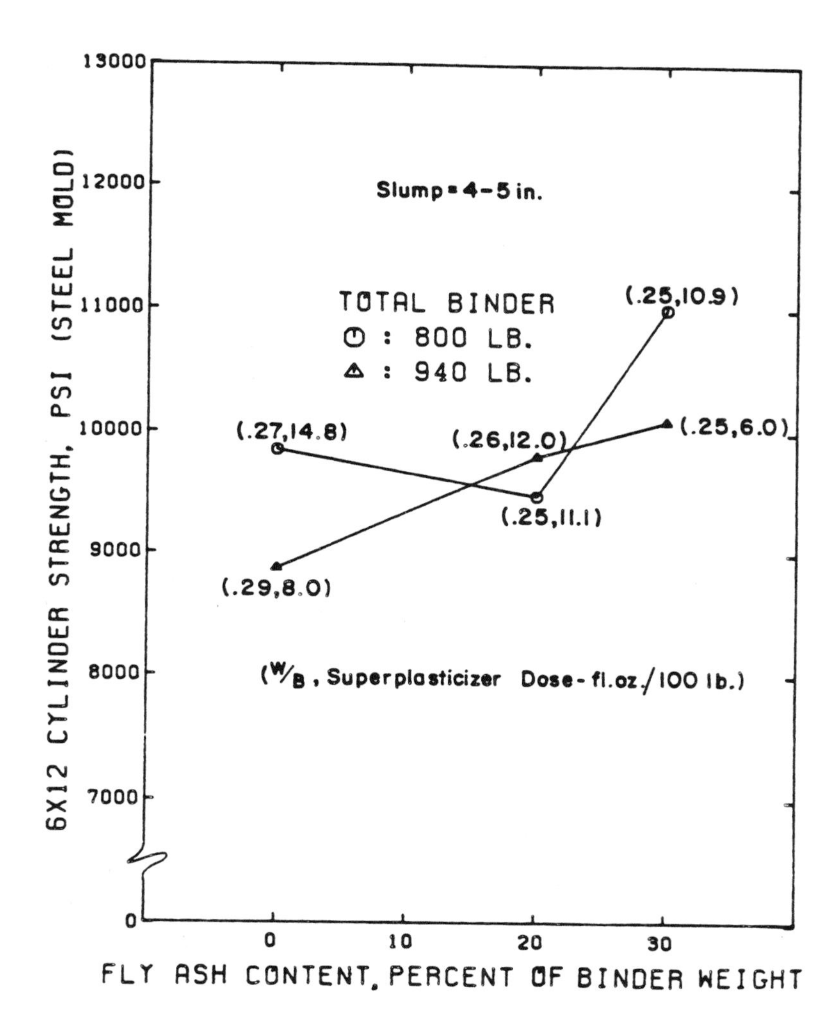

Fig. 4.76 Effect of total binder content and fly ash content on the 28-day compressive strength of concrete for mixes having a CA/FA ratio of 2.0 and made with type II cement, fly ash A, 1/2-in. limestone E, sand B, and superplasticizer B.

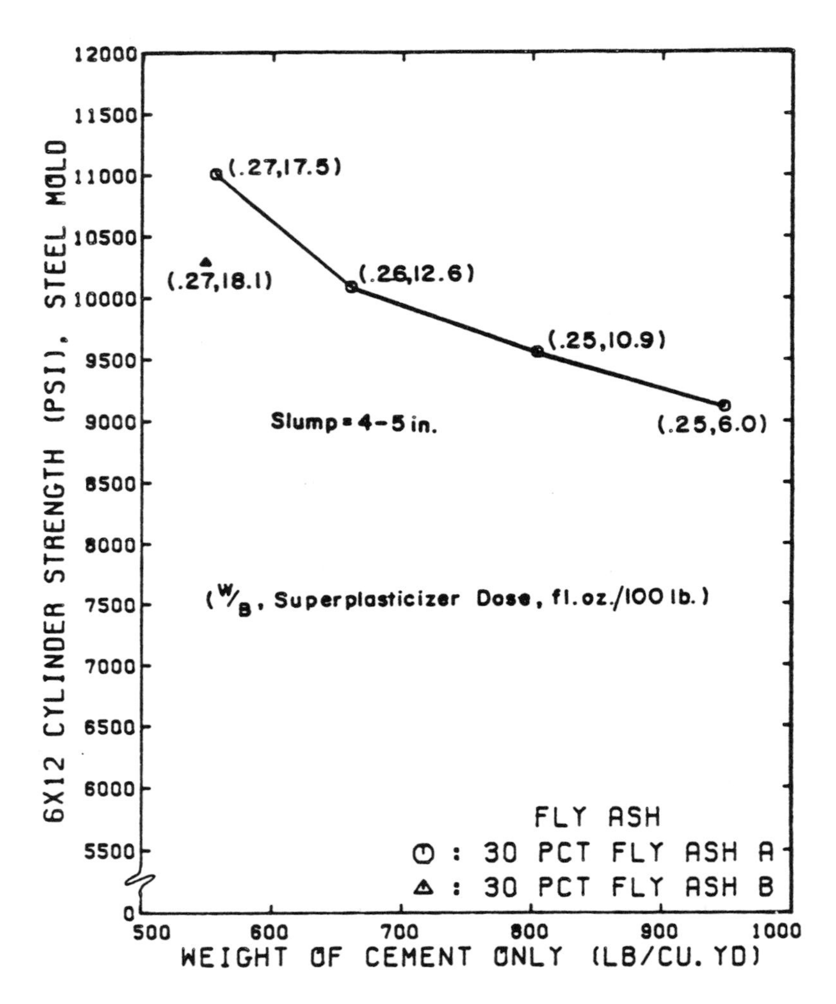

Fig. 4.77 Effect of fly ash source and cement content on the 28-day compressive strength of concrete for mixes having a CA/FA ratio of 2.0 and made with type II cement, 1/2-in. limestone E, sand B, and superplasticizer B.

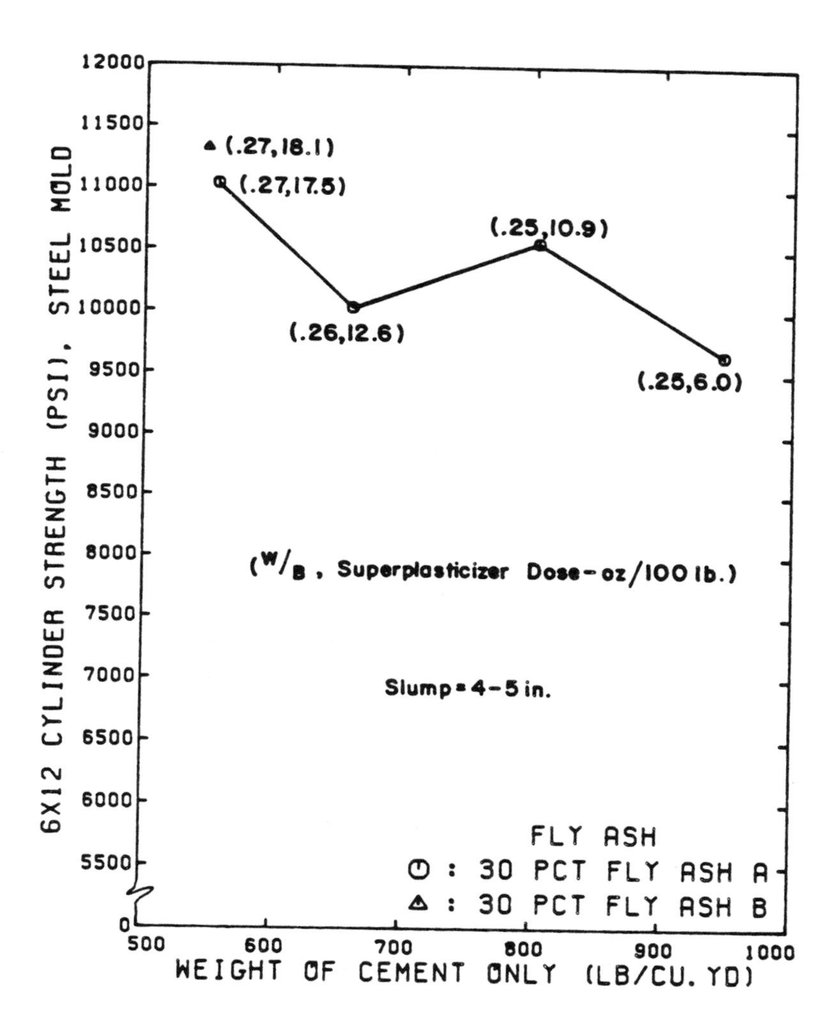

Fig. 4.78 Effect of fly ash source and cement content on the 56-day compressive strength of concrete for mixes having a CA/FA ratio of 2.0 and made with type II cement, 1/2-in. limestone E, sand B, and superplasticizer B.

4.10.3 Fly Ash and Coarse/Fine Aggregate Ratio. The effect of very high coarse aggregate contents on compressive strength was studied for mixes containing approximately 12 sacks of binder/cu.yd. and a fly ash content of 20 percent of total binder weight. As shown in Figs. 4.79 and 4.80, the compressive strength of concrete decreased with an increased CA/FA ratio. The loss of strength with increasing CA/FA was greater for mixes containing superplasticizer.

4.11 Effect of Temperature and Mixing Time

The effects of high temperature and mixing time on slump and compressive strength of high strength concrete were studied. High strength concrete mixes made with and without superplasticizers, fly ash, and reducer-retarders were considered. Some mixes made with and without superplasticizers were retempered with superplasticizer and water, respectively, to adjust the slump after mixing for prolonged periods at temperatures of approximately 100°F.

Tables 4.3 through 4.6 list mix proportions and compressive strength data for four different sets of high strength concrete mixes. Slump is plotted versus mixing time in Figs. 4.81 through 4.84 for these mixes.

Slump losses after 60 minutes of mixing at high temperature ranged from 0 in. to only 1-3/4 in. for concrete containing no fly ash. For similar mixes containing fly ash, slump dropped to 0 in. after 60 minutes. Slump loss rates were similar after retempering, with or without chemical admixtures.

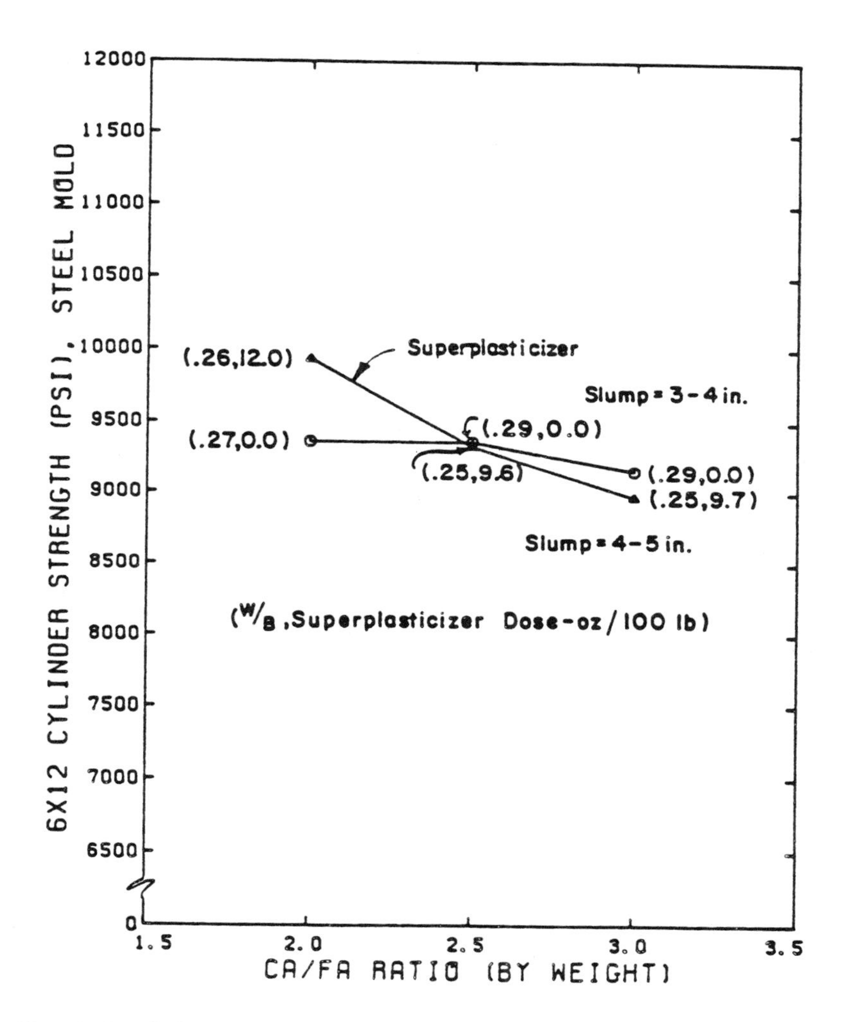

Fig. 4.79 Effect of high CA/FA ratios and superplasticizer on the 28-day compressive strength of concrete for mixes having a cement content of 10 sacks/cu.yd. and made with type II cement, fly ash A(20% by wt.), 1/2-in. limestone E, and sand B.

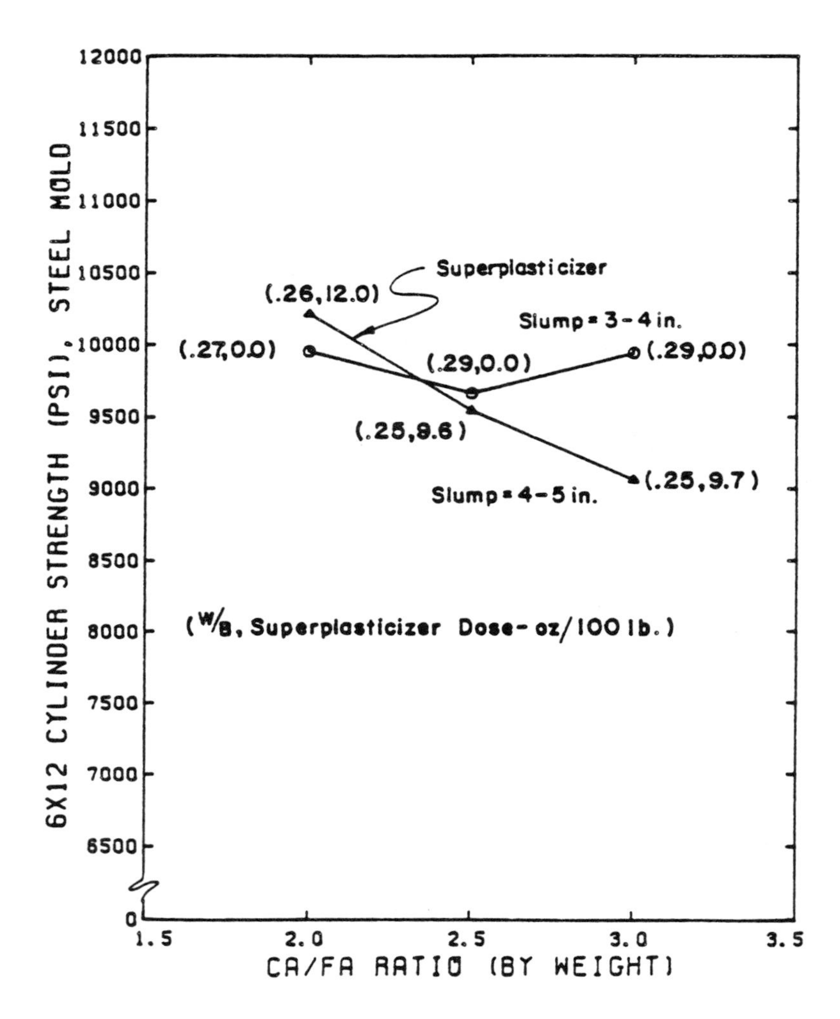

Fig. 4.80 Effect of high CA/FA ratios and superplasticizer on the 56-day compressive strength of concrete for mixes having a cement content of 10 sacks/cu.yd. and made with type II cement, fly ash A (20% by wt.), 1/2-in. limestone E, and sand B.

TABLE 4.3 Mix Design Data per Cu. Yd. for 10-Sack High Strength Concrete Mix Containing No Superplasticizer and No Fly Ash (Mix Q).

	Reference Mix Q	Mix Q-1	Mix Q-2a	Mix Q-2b
Cement C (lbs)	930	943	933	930
1/2-in. Limestone E (lbs)	1825	1804	1785	1780
Sand B (lbs)	901	902	893	890
Reducer-Retarder C (fluid ounces)	0	0	45	45
Initial W/C Ratio	—	.35	.34	.34
Final W/C Ratio (a)	.35	.35	.36	.37
Mixing Temperature (°F)	72	106	102	103
Mixing Time (min.) (b)	15	60	60	90
Compressive Strength (c) (psi)	8890	7930	9470	9050

[a] Refers to the water-cement ratio at time of casting of cylinders. All retempering water added to restore the workability of the mix is included as part of the water.

[b] Refers to the duration of the mixing until casting of compression cylinders.

[c] Refer to the 6-in. dia. x 12-in. cylinder compressive strength of specimens cast from that mix moist cured, and tested at 28 days. (Average of three specimens)

TABLE 4.4 Mix Design Data per Cu. Yd. for 8.5-Sack High Strength Concrete Mix Containing Superplasticizer but No Fly Ash (Mix R).

	Reference Mix R	Mix R-1a	Mix R-1b	Mix (d) R-2a	Mix (d) R-2b
Cement C (lbs)	785	798	797	803	802
1/2-in. Limestone E (lbs)	2041	2065	2064	2079	2075
Sand B (lbs)	1011	1034	1033	1040	1038
Superplasticizer B Initial Dose (fl.oz.)	---	129	129	126	126
Final Dose (fl.oz.)	124	129	153	126	180
Reducer-Retarder D (fluid ounces)	0	0	0	43	43
Water/Cement Ratio (a)	.32	.28	.28	.27	.27
Mixing Temperature (°F)	71	104	97	104	99
Mixing Time (min.) (b)	15	60	90	60	90
Compressive Strength (c) (psi)	10,610	10,400	11,470	11,490	11,820

[a] Refers to the water-cement ratio at time of casting of cylinders. All admixture added to restore the workability of the mix is included as part of the water.

[b] Refers to the duration of the mixing until casting of compression cylinders.

[c] Refers to the 6-in. dia. x 12-in. cylinder compressive strength of specimens cast from that mix, moist cured, and tested at 28 days. (Average of three specimens)

[d] Demolded 48 hours after casting due to slow rate of hardening.

TABLE 4.5 Mix Design Data per Cu. Yd. for 7-Sack High Strength Concrete Mix Containing 30% Fly Ash but No Superplasticizer (Mix S).

	Reference Mix S	Mix S-1a	Mix S-1b	Mix S-2a	Mix S-2b
Cement C (lbs)	653	646	638	662	654
Fly Ash A (lbs) (Class C)	280	276	273	283	280
1/2-in. Limestone E (lbs)	1821	1803	1782	1848	1826
Sand B (lbs)	916	901	891	923	913
Reducer-Retarder C (fluid ounces)	0	0	0	35	35
Initial W/C Ratio	—	.46	.46	.38	.38
Final W/C Ratio (a)	.46	.49	.52	.43	.46
Initial W/B Ratio	—	.32	.32	.26	.26
Final W/B Ratio (a)	.32	.34	.36	.30	.32
Mixing Temperature (°F)	70.5	108	108	101	101
Mixing Time (min.) (b)	15	60	90	60	90
Compressive Strength (c) (psi)	9630	8490	8080	9650	9590

[a] Refers to the water-cement and water-binder ratios at time of casting of cylinders. All retempering water added to restore the workability of the mix is included as part of the water.

[b] Refers to the duration of the mixing until casting of compression cylinders.

[c] Refers to the 6-in. dia. x 12-in. cylinder compressive strength of specimens cast from that mix, moist cured, and tested at 28 days. (Average of three specimens)

TABLE 4.6 Mix Design Data per Cu. Yd. for 6-Sack High Strength Concrete Mix Containing Superplasticizer and 30% Fly Ash (Mix T)

	Reference Mix T	Mix T-1a	Mix T-1b	Mix (d) T-2a	Mix (d) T-2b
Cement C (lbs)	553	565	564	566	565
Fly Ash A (lbs) (Class C)	237	243	242	243	243
1/2-in. Limestone E (lbs)	2,042	2,072	2,069	2,078	2,073
Sand B (lbs)	1,040	1,036	1,035	1,039	1,036
Superplasticizer B					
Initial Dose (fl.oz)	---	108	108	109	109
Final Dose (fl.oz.)	135	153	184	148	211
Reducer-Retarder D (fluid ounces)	0	0	0	30	30
Water/Cement Ratio (a)	.40	.36	.36	.35	.36
Water/Binder Ratio (a)	.28	.25	.25	.24	.25
Mixing Temperature (°F)	72.5	104	102	105	105
Mixing Time (min.) (b)	15	60	90	60	90
Compressive Strength (c) (psi)	11,600	11,210	11,430	12,170	12,160

[a] Refers to the water-cement and water-binder ratios at time of casting of cylinders. All retempering admixture added to restore the workability of the mix is included as part of the water.

[b] Refers to the duration of the mixing until casting of compression cyliners.

[c] Refers to the 6-in. dia. x 12-in. cylinder compressive strength of specimens cast from that mix, moist cured, and tested at 28 days. (Average of three specimens)

[d] Demolded 48 hours after casting due to slow rate of hardening.

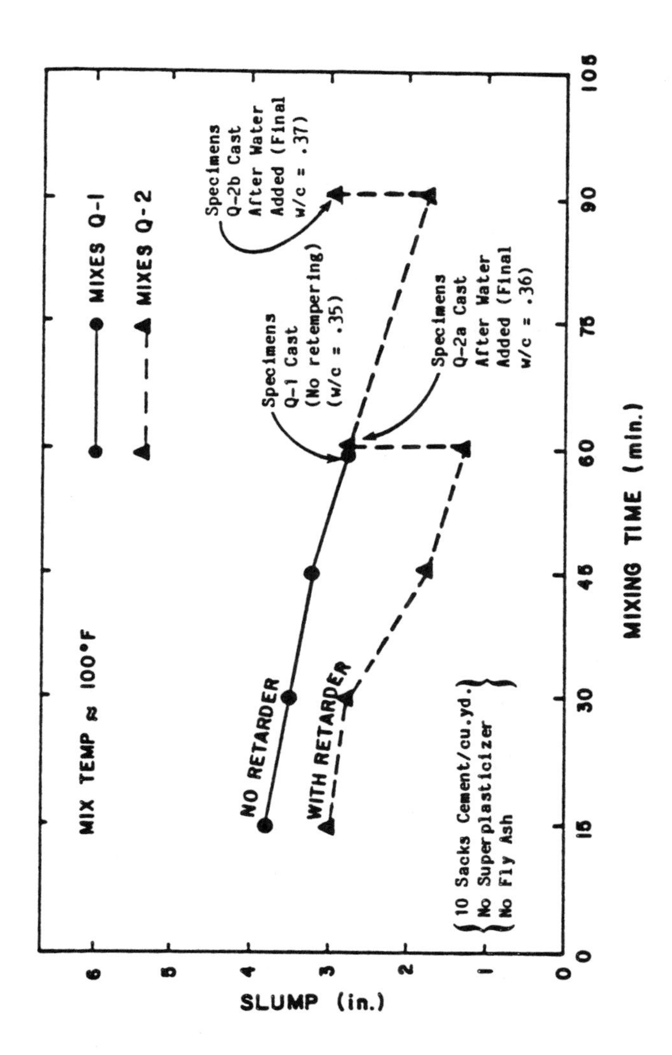

Fig. 4.81 Effect of mixing time and a water-reducing-retarding admixture on the slump of high strength concrete mixes having a cement content of 10 sacks/cu.yd. but containing no fly ash or superplasticizer (see Table 4.3).

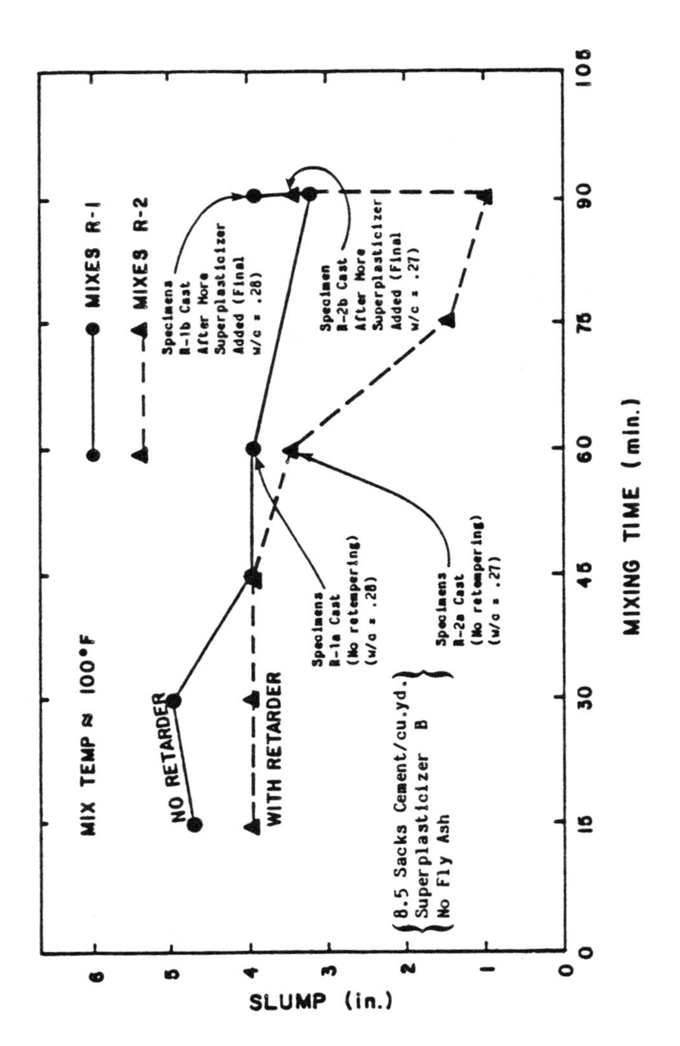

Fig. 4.82 Effect of mixing time and a water-reducing-retarding admixture on the slump of high strength concrete mixes having a cement content of 8.5 sacks/cu.yd. and containing a superplasticizer but no fly ash (see Table 4.4).

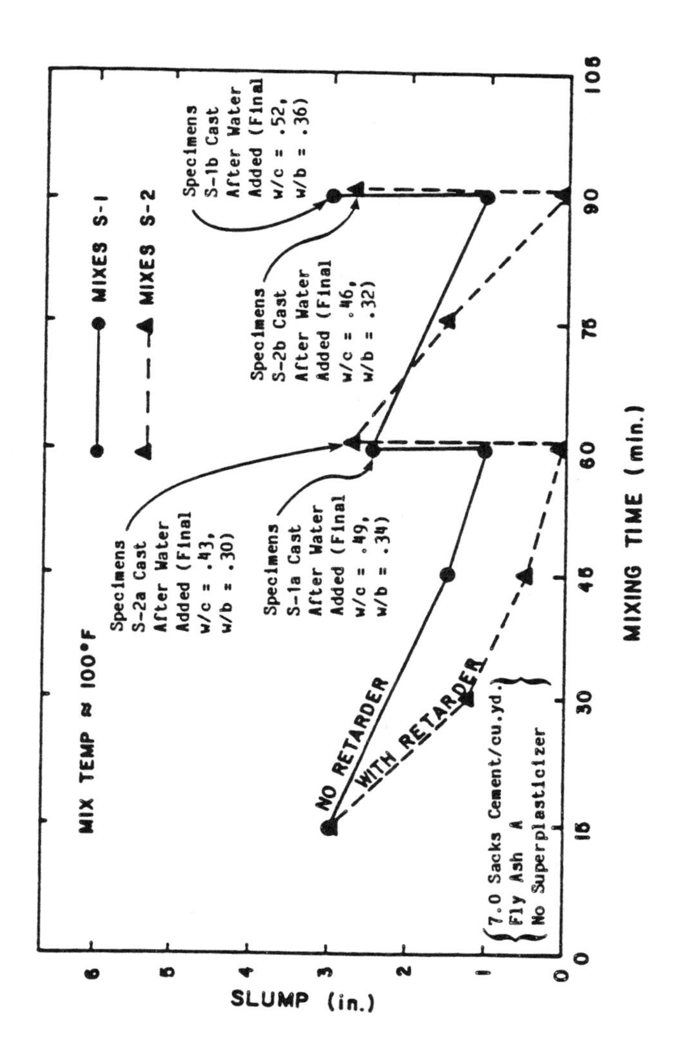

Fig. 4.83 Effect of mixing time and a water-reducing-retarding admixture on the slump of high strength concrete mixes having a cement content of 7.0 sacks/cu.yd. and containing fly ash but no superplasticizer (see Table 4.5).

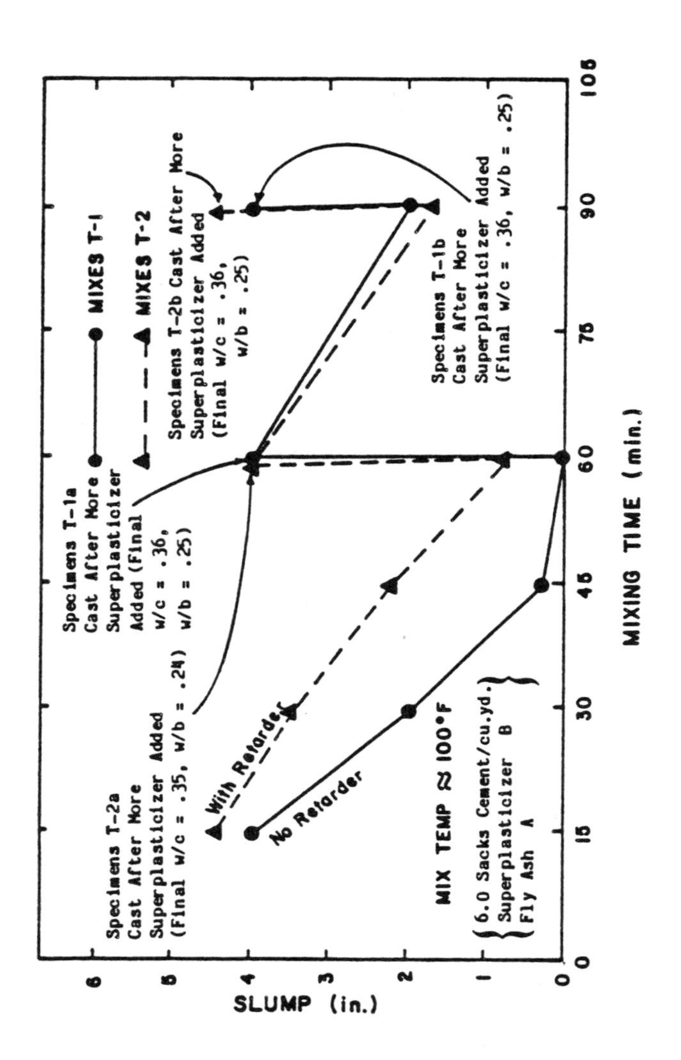

Fig. 4.84 Effect of mixing time and a water-reducing-retarding admixture on the slump of high strength concrete mixes having a cement content of 6.0 sacks/cu.yd. and containing fly ash and superplasticizer (see Table 4.6).

As expected, compressive strength of high strength concrete was reduced when water was added to mixes containing no chemical admixture to restore slump after 60 to 90 min. of mixing at 100°F, as shown in Tables 4.3 and 4.5. Adding superplasticizer instead of water to restore slump in mixes already containing superplasticizer resulted in a significant increase in compressive strength, as shown in Tables 4.4 and 4.6. The addition of water-reducer-retarders generally did not affect the rate of slump loss with time in the mixes studied. However, for all mixes, the addition of a reducer-retarder admixture to a mix batched at or above 100°F resulted in a higher 28-day compressive strength than that of the same basic concrete mix batched at 70°F to 75°F and containing no reducer-retarder admixture.

Reducer-retarder C was used at a dosage of 2.0 fl.oz./100 lbs of cement in the two mixes containing no superplasticizer. The rate of slump loss was not improved by the addition of water-reducer C to these mixes at this dosage. As a result, a second reducer-retarder D was used instead in the remaining mixes containing superplasticizer. Reducer-retarder D was added at a dosage rate of 5.0 fl.oz./100 lbs of cement to the mixes containing superplasticizers. This admixture dosage was well within the manufacturer's recommended dosage. However, hardening of the fresh concrete was retarded so much that specimens could not be demolded 24 hours after casting. These specimens were demolded 48 hours after casting.

When the mix proportions of the high temperature batch labeled "T-2a" in Table 4.6 were remixed with a reducer-retarder dosage of only

2.0 fl.oz./100 lbs of cement, the specimens were demolded 24 hours after casting without problems. The compressive strength at 24 hours was 4690 psi. Also, when the same mix proportions were batched at 78°F with a reducer-retarder dosage of 4.0 fl.oz./100 lbs, the specimens were ready for demolding 24 hours after casting.

4.12 High Strength Concrete and Test Age

Four different high strength concrete mixes were tested for compressive strength at curing ages of 1, 7, 28 and 56 days. The mix proportions for these concretes are listed in Table 4.7. Test results are shown in Fig. 4.85.

The 1-day strength of both mixes containing superplasticizer with and without fly ash was of the order of 6,000 psi. For mixes containing no superplasticizers, the 1-day compressive strength was approximately 4,200 psi. The addition of 30 percent fly ash to mixes with and without superplasticizers reduced slightly the 1-day compressive strength compared to similar mixes containing no fly ash. At later ages mixes containing fly ash showed a higher compressive strength than similar mixes containing no fly ash. For all mixes, compressive strengths at 28 days ranged from about 9,000 psi for a plain 10-sack mix containing no chemical or mineral admixtures to about 11,500 psi for a mix containing 6 sacks of cement per cu.yd., superplasticizer, and fly ash.

When a water-reducer-retarder was added to the 6-sack mix containing superplasticizer and fly ash, made at room temperature, the compressive strength at 24 hours was reduced from 5,900 psi to 4,700

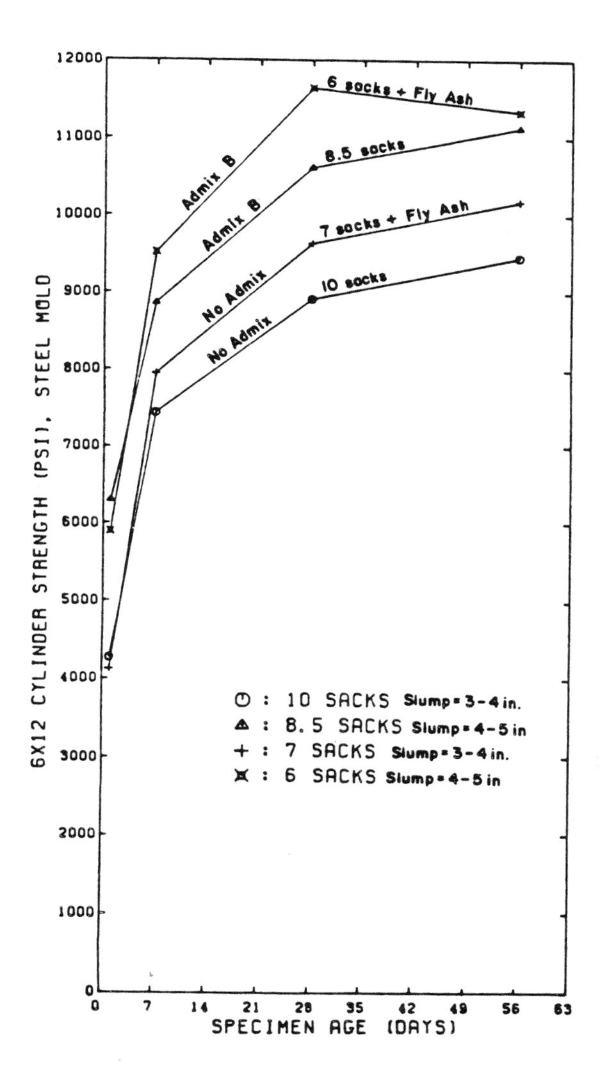

Fig. 4.85 Effect of specimen age and fly ash and a superplasticizer on the compressive strength of concrete for mixes having a CA/FA ratio of 2.0 and made with type II cement, fly ash A (0% or 30% by wt.), 1/2-in. limestone E, and sand B.

TABLE 4.7 Mix Proportions for Concrete Mixes Shown in Fig. 4.85, Comparing Compressive Strengths at Different Curing Ages.

	Mix "Q" 10 Sacks	Mix "R" 8.5 Sacks	Mix "S" 7.0 Sacks	Mix "T" 6.0 Sacks
Cement C (lbs)	921	785	653	553
Fly Ash A (lbs) (Class C)	0	0	280	237
1/2-in. Limestone E (lbs)	1,834	2,041	1,821	2,042
Sand B (lbs)	866	1,011	916	1,040
Superplasticizer B (fluid ounces)	0	117	0	127
Water/Cement Ratio	.37	.32	.46	.40
Water/Binder Ratio	.37	.32	.32	.28
28-day f'_c (psi)	8,910	10,610	9,630	11,640

psi. However, as can be seen in Fig. 4.86, the 7-day and 28-day compressive strengths of the retarded mix were equivalent to those of the same mix without the retarder.

4.13 Compaction, Curing and Capping

The effects of different compaction, curing, and capping procedures on 28-day compressive strength of high strength concrete are compared in Table 4.8. The highest 28-day compressive strength was achieved by moist curing for 14 days followed by oven drying at 100°F to 120°F. Cylinders compacted by rodding resulted in higher compressive strengths than cylinders compacted by 2 min. of external vibration. Using high strength capping compound material results in higher concrete compressive strength test results than using conventional sulfur compounds.

These results show that the compressive strength of high strength concrete is not adversely affected by a hot and dry environment after 7 days of ideal curing.

4.14 Flexural Strength

Third-point loading, flexural beam tests at 28 days were performed for most concrete mixes in this project.

As shown in Fig. 4.87, the flexural strength of all mixes tested fell within the range from $8.0\sqrt{f'_c}$ to $12.0\sqrt{f'_c}$.

4.15 Split Cylinder Strength

Split cylinder tests were performed on 6-in. x 12-in. cylinders from several mixes. Split cylinder tensile strength results were

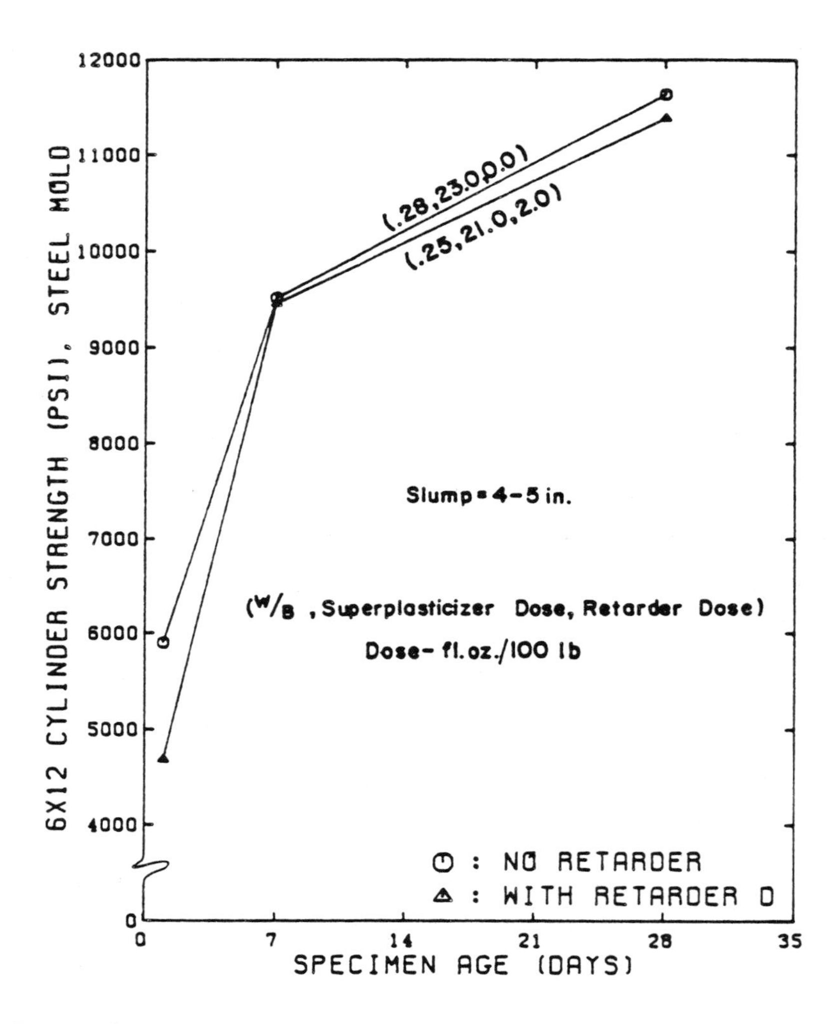

Fig. 4.86 Effect of specimen age and water-reducing-retarding admixture D on the compressive strength of concrete for mixes having a cement content of 6 sacks/cu.yd. and a CA/FA ratio of 2.0 and made with type II cement, fly ash A (30% by wt.), 1/2-in. limestone E, sand B, and superplasticizer B.

TABLE 4.8 Effects of Different Compaction, Curing, and Capping Procedures on 28-Day Compressive Strength. (4-in. Dia. x 8 in. Cylinders, Cardboard Molds)

Curing	f'_c (psi)
28 days, under water	11,050
28 days, moist, 73°F	10,550
14 days, moist, 73°F 14 days, dry, 80°F-90°F	11,480
7 days, moist, 73°F 21 days, dry, 80°F-90°F	11,380
14 days, moist, 73°F 14 days, dry, 100°F-120°F	12,360
7 days, moist, 73°F 21 days, dry, 100°F-120°F	12,260

Compaction	f'_c (psi)
Rodding	10,550
External Vibrating, 2 min.	10,170

Capping	f'_c (psi)
Sulphur Capping Material A	10,550
Sulphur Capping Material B	11,180

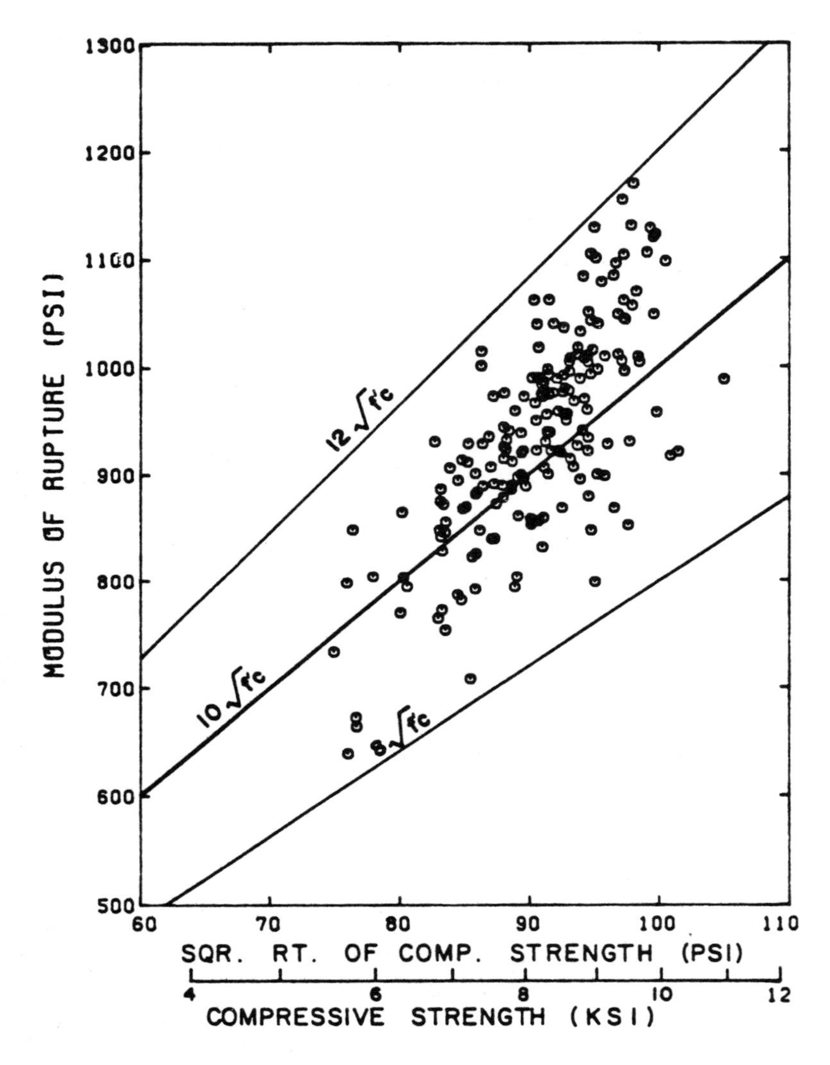

Fig. 4.87 Plot of modulus of rupture versus compressive strength for all flexure test beam specimens made in this study. Each point represents the average of three test results.

approximately $8.1\sqrt{f'_c}$ for specimens molded in steel forms, and $7.7\sqrt{f'_c}$ for those made using cardboard molds. Identical mixes tested for flexural strength had an average modulus of rupture of $10.4\sqrt{f'_c}$.

4.16 Mold Types and Sizes

The effects on compressive strength of high strength concrete of using cylindrical concrete specimen molds made of steel, plastic, and cardboard were compared. The effects of specimen size on compressive strength were studied as well. Based on the test results using 4-in. dia. x 8-in. cylinder specimens, concrete made in steel molds always had higher compressive strength, than specimens made using cardboard, as seen in Fig.4.88. Table 4.9 shows the results of four high strength concrete mixes made to compare 6-in. dia. x 12-in. cylinders made out of cardboard, plastic and steel molds. Specimens made in steel molds were generally stronger than those made in cardboard molds. No definite conclusions can be made from this data with respect to strength of specimens made using plastic molds as compared to steel molds.

As seen in Fig. 4.89, 4-in. dia. x 8-in. cylinders always gave higher compressive strength results than 6-in. dia. x 12-in. specimens when cast in molds made of the same material. Generally, a 4-in. x 8-in. cylinder made using steel molds can be expected to result in a compressive strength of between 10 and 15 percent higher than a specimen made out of the same batch using a 6-in. x 12-in. steel mold.

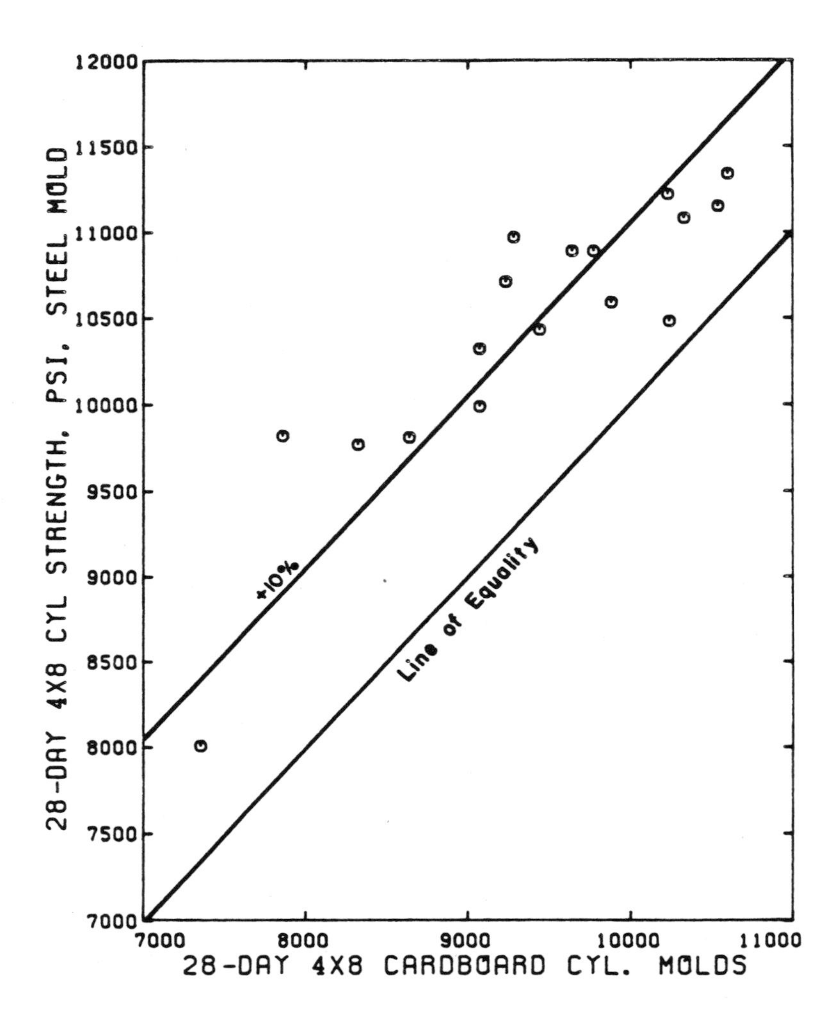

Fig. 4.88 Comparison of the 28-day compressive strength of high strength concrete specimens cast in 4-in. dia. x 8-in. cardboard and rigid steel molds.

TABLE 4.9 Compressive Strength Test Results of Specimens Cast in Different Types of Molds.

Mold Material	Cylinder Size	28-Day f'_c (psi)			
		Mix Q	Mix R	Mix S	Mix T
		(Plain)	(Admix, No Fly Ash)	(Fly Ash, No Admix)	(Admix, & Fly Ash)
Steel	6-in. dia. x 12-in.	8,890	9,500	9,560	10,210
Cardboard	6-in. dia. x 12-in.	8,490	9,730	9,090	10,060
Plastic	6-in. dia. x 12-in.	8,230	10,730	8,930	10,960
Steel	4-in. x 8-in.	9,810	11,150	10,480	11,080
Cardboard	4-in. x 8-in.	8,640	10,540	10,240	10,330

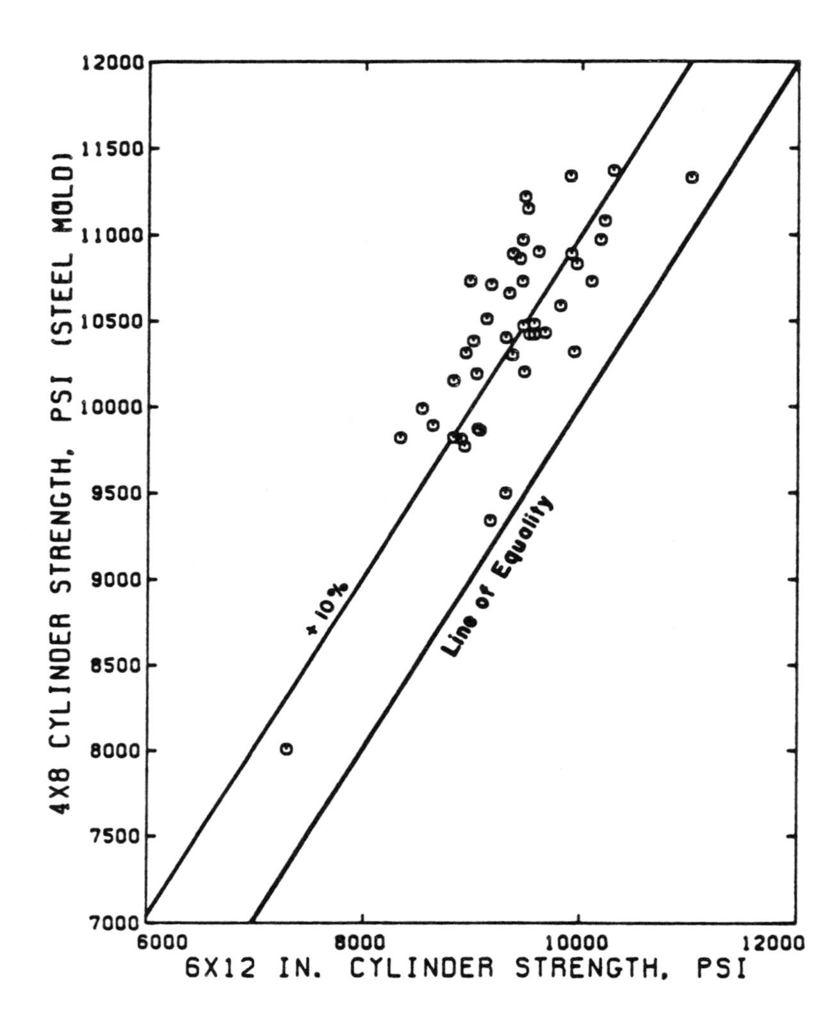

Fig. 4.89 Comparison of the 28-day compressive strength of high strength concrete specimens cast in 4-in. dia. x 8-in. and 6-in. dia. x 12-in. rigid steel molds.

4.17 Superplasticizers and Workability

When superplasticized mixes were first introduced into this program, it was intended for all batches to contain a final water-cement ratio of 0.30 or less. However, 7-sack mixes containing superplasticizer, 1/2-in. coarse aggergate, and coarse/fine ratios of 1.0 to 2.0 required a superplasticizer dose of more than 50 fl.oz./100 lbs of cement in order to achieve a 4-in. slump at a water-cement ratio of 0.30 or less. As a result of this high dosage, excessive bleeding occurred and the coarse aggregate had a slimy appearance. The fresh concrete had no cohesion, and workability requirements were not adequate. When attempting to measure the slump of this concrete, the sample slowly collapsed to nearly a 12-in. "slump" after removal of the slump cone. The mix was rocky and too harsh to rod and compact properly. The specimens were soft and crumbly 24 hours after casting and flexure beam specimens would fail under their own weight when supported at an 18-in. span. One of these 6-in. dia. x 12-in. cylinder specimens was saved and demolded at 3 days. Its 3-day compressive strength was 4000 psi. Its appearance was dark brown and porous.

Reducing the superplasticizer dosage from 50 fl.oz./100 lbs of cement to 25 fl.oz./100 lbs of cement was insufficient to make possible a workable mix containing 7 sacks of cement per cu.yd. and having a w/c ratio of 0.30. The 7-sack mix with a dosage of 25.0 fl.oz./100 lbs of cement was too harsh and the slump collapsed. The specimens were not ready to be demolded for 48 hours after casting. The 56-day compressive strength (6,500 psi) of this concrete was only 2/3 of that obtained with the same mix to which a superplasticizer dose of 15 fl.oz./100 lbs of cement had been added.

A superplasticizer dosage of 15 fl.oz./100 lbs of cement in a mix which had a w/c ratio of at least 0.33 was the maximum acceptable dosage for workability in a 7-sack mix using the materials in this study. A lower w/c ratio can be obtained with a higher admixture dosage, but workability and strength are sacrificed.

For 8.5- and 10-sack mixes the higher fines contents allowed superplasticizer doses greater than 30 fl.oz./100 lbs cement to be added to the concrete without workability problems. Slump test results were more representative of the workability of these mixes.

Figure 4.90 is a diagram relating the workability of mixes in this study which contained superplasticizers, a 1/2-in. crushed limestone coarse aggregate, and type II cement, and had a slump of 4 in. to 5 in. At one extreme, a lean (7-sack) mix with a high coarse aggregate content (coarse/fine aggregate ratio = 2.0) was harsh and unworkable with a strong tendency to segregate. At the other extreme, a rich, 10-sack mix with the highest fine aggregate content (coarse/fine ratio = 1.0) was too sticky.

The "slightly rocky" mixes frequently appeared to stiffen when sitting still in molds or in the mixer. However, the concrete quickly loosened and flowed when remixed or subjected to vibration, especially internal vibration.

Changes in the materials used affected workability. For instance, 7-sack mixes having a CA/FA ratio of 2.0 were workable when

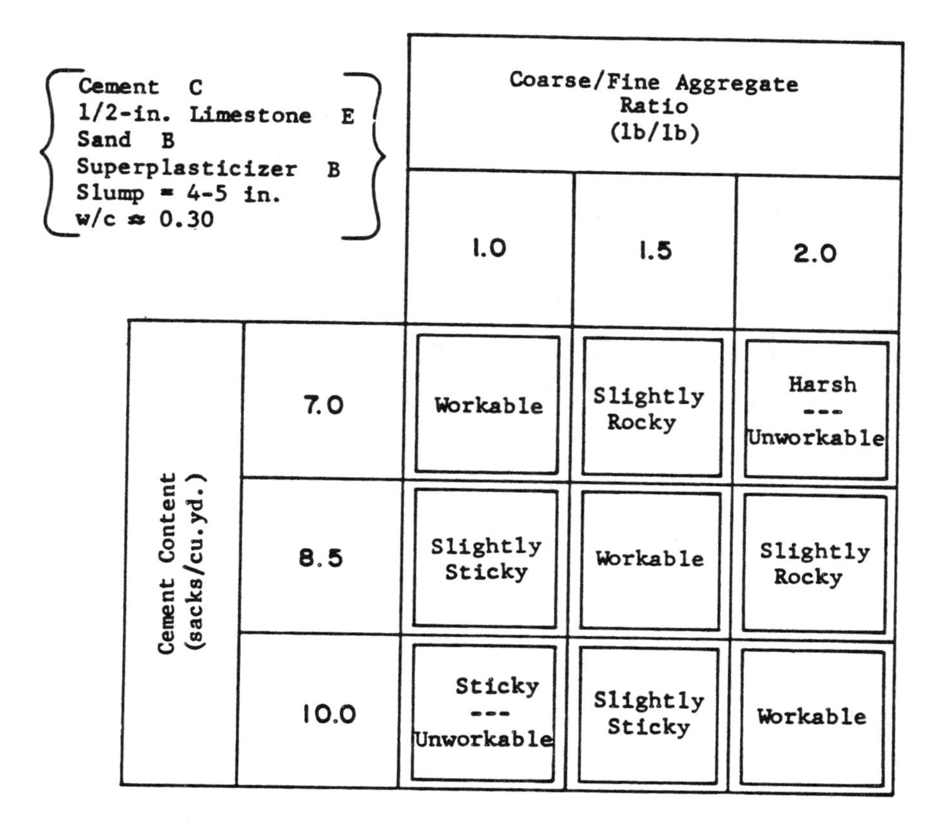

Fig. 4.90 Effect of the CA/FA ratio and cement content on the workability of concrete mixes made with 1/2-in. limestone coarse aggregate and a superplasticizer.

made with 3/4-in. stone instead of 1/2-in. stone. Brand of superplasticizer also affected workability. More bleeding and segregation occurred in 7-sack mixes made using superplasticizer brand A than in mixes made with brand B.

Due to the higher fines content of high strength concrete, formed concrete surfaces resulted in a satisfactory appearance. However, in this study, hand-finishing was difficult on the top surface of specimens made from "slightly rocky" mixes containing superplasticizers and especially difficult for the "harsh" mixes. This is not expected to present significant problems in casting columns and precast girders in the field, since use of power finishers has reportedly resulted in acceptable finished surfaces [90].

V. Discussion of Test Results

5.1 Introduction

The experimental test results presented in Chapter IV are discussed in this chapter. Explanations for the observed effects of different variables on compressive strength of high strength concrete are examined. Procedures are suggested for direct application of the test results presented in Chapter IV to the development of high strength concrete mix designs in concrete batching plants in the state of Texas.

5.2 Cement Content

In order to produce high strength concrete, higher cement contents than for normal strength concrete must be used, as shown in Fig. 5.1. The cement content of concrete mixes made in this study ranged from 7.0 sacks/cu.yd. to 10.0 sacks/cu.yd. For trial mix design programs in Texas, cement contents in excess of 8.5 sacks/cu.yd. and as high as 11.0 or 12.0 sacks/cu.yd. should be used for concrete mixes containing no fly ash and no chemical admixtures. When evaluating the effects of cement content and superplasticizer dosage on concrete strength, cement contents in the range from 6.0 to 10.0 sacks/cu.yd. should be considered. However, the workability of mixes containing superplasticizers and having a cement content ranging from 6.0 to 8.0 sacks/cu.yd., and a coarse/fine aggregate ratio (CA/FA) of 1.5 or

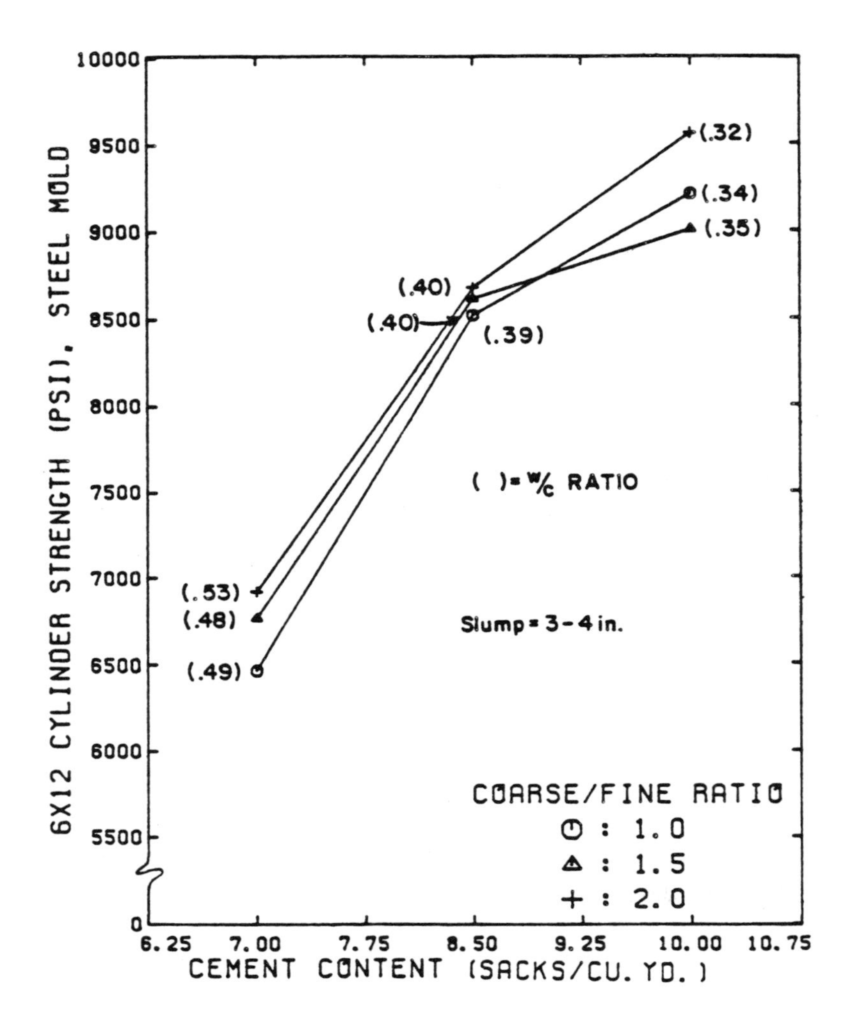

Fig. 5.1 Effect of cement content and CA/FA ratio on the 56-day compressive strength of concrete for mixes made with type II cement, 1/2-in. limestone E, sand B, and no admixture.

greater might not be acceptable for placement of the concrete in the field.

Ten-sack mixes containing no superplasticizer or fly ash produced 56-day compressive strengths greater than 9,000 psi using 1/2-in. max. size coarse aggregate. Two 8.5-sack mixes having a CA/FA ratio of 1.5 and made with identical materials except for the source of the sand resulted in concrete strengths of 9,000 psi at 56 days. However, no other 7.0 or 8.5-sack mixes resulted in 9,000 psi concrete at 56 days without the use of chemical or mineral admixtures. The low mixing water requirement associated with high cement factors was greatly responsible for achieving high strength in mixes containing no chemical or mineral admixtures. Typically, the w/c ratio for a 1/2-in. max. size crushed stone, 10-sack mix having a compressive strength of 9000 psi at 56 days containing no admixture was 0.32 for a 3-in. slump. However, for a similar 8.5-sack mix, the w/c ratio was about 0.37. To produce concrete having a 3-in. slump, mixes containing 7.0 sacks/cu.yd. required a w/c ratio of between 0.42 and 0.50.

For mixes containing no superplasticizer, increasing the cement content from 8.5 to 10.0 sacks/cu.yd. in mixes made using 1/2-in. max. size coarse aggregate resulted in significant increases in compressive strength at any age. For mixes containing no admixtures, using cement contents in excess of 10 sacks/cu.yd. may result in even higher compressive concrete strengths than were obtained without admixtures in this study. However, all mixes made in this study had cement contents of 10 sacks/cu.yd. or less.

When superplasticizers were added to high strength concrete mixes, increases in compressive strength of the concrete were observed at test ages of 28 and 56 days for any cement content. For producing high strength concrete, the optimum cement content for mixes containing superplasticizer was 8.5 sacks/cu.yd., as shown in Fig. 5.2. The dispersion effect of the superplasticizer on cement particles improved the efficiency of hydration, making the strength of the concrete less dependent upon the cement content and w/c ratio. The higher the superplasticizer dosage, the higher the compressive strength of the concrete for a given workability as long as the mix remained cohesive. Since the objective of using superplasticizers was to reduce the w/c ratio of all mixes to 0.30 while maintaining a 4-in. slump, the 8.5-sack mixes required higher admixture dosages than did the 10.0-sack mixes, due to the much higher w/c ratio of the 8.5-sack mixes without superplasticizer. The 7.0 sack mixes required the highest admixture dosages of all but, as reported in Section 4.17 on superplasticizers and workability, in order to produce a good, workable concrete having a normal setting time and a smooth formed surface, the w/c ratio of 7.0-sack mixes had to be increased to at least 0.34. As a result, 8.5-sack mixes containing superplasticizer in this study produced the highest compressive strengths. It may be possible to obtain higher concrete strengths having the desired 4 to 5-in. slump with much lower w/c ratios through the use of higher superplasticizer dosages. However, no attempt was made in this study to obtain the lowest possible w/c ratio.

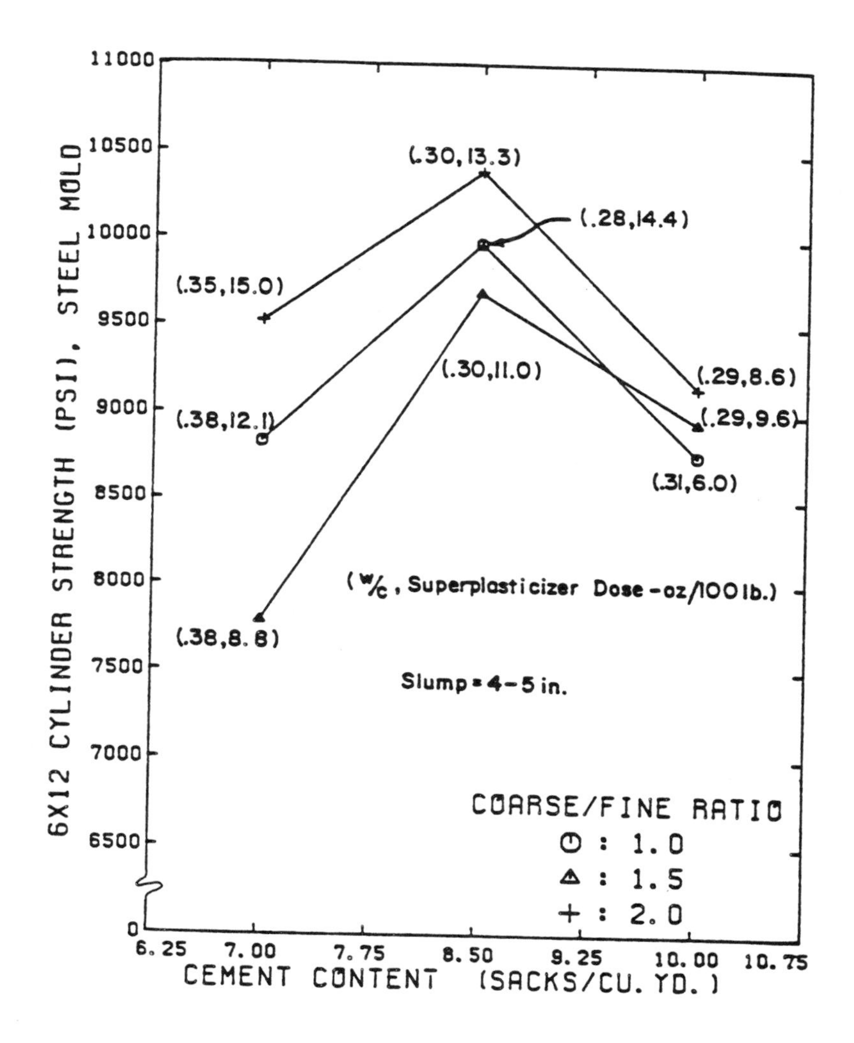

Fig. 5.2 Effect of cement content and CA/FA ratio on the 56-day compressive strength of concrete for mixes made with type II cement, 1/2-in. limestone E, sand C, and superplasticizer B.

For mixes made using superplasticizers, compaction of specimens was most effective in the 8.5-sack mixes, and may have also enhanced the compressive strength of those mixes. Fresh concrete mixes containing 7.0 sacks of cement per cu.yd. tended to be harsh while the 10-sack concrete mixes were generally sticky for optimum compactibility using current cylinder casting techniques.

5.3 Water/Cement Ratio

Lower water/cement ratios are required for producing high strength concrete than for producing normal strength concrete. In addition, use of the water-binder ratio is more appropriate than use of the water-cement ratio as a general indicator of the compressive strength of high strength concrete. The plot of concrete compressive strength versus water-binder (w/b) ratio shown in Fig. 5.3 indicates that a w/b ratio of less than about 0.32 is required for producing concretes having a 28-day compressive strength of 9,000 psi. A w/b ratio less than about 0.35 is required for producing concretes having a 56-day strength of 9,000 psi or higher. This is based on a minimum 4-in. slump for mixes containing superplasticizer and a minimum slump of 3 in. for mixes containing no superplasticizer. Higher strengths would be produced if the slump were permitted to be less than 3 to 4 in., since the w/c or w/b ratio could be reduced. However, all concrete mixes produced in this study had slumps of at least 3 in.

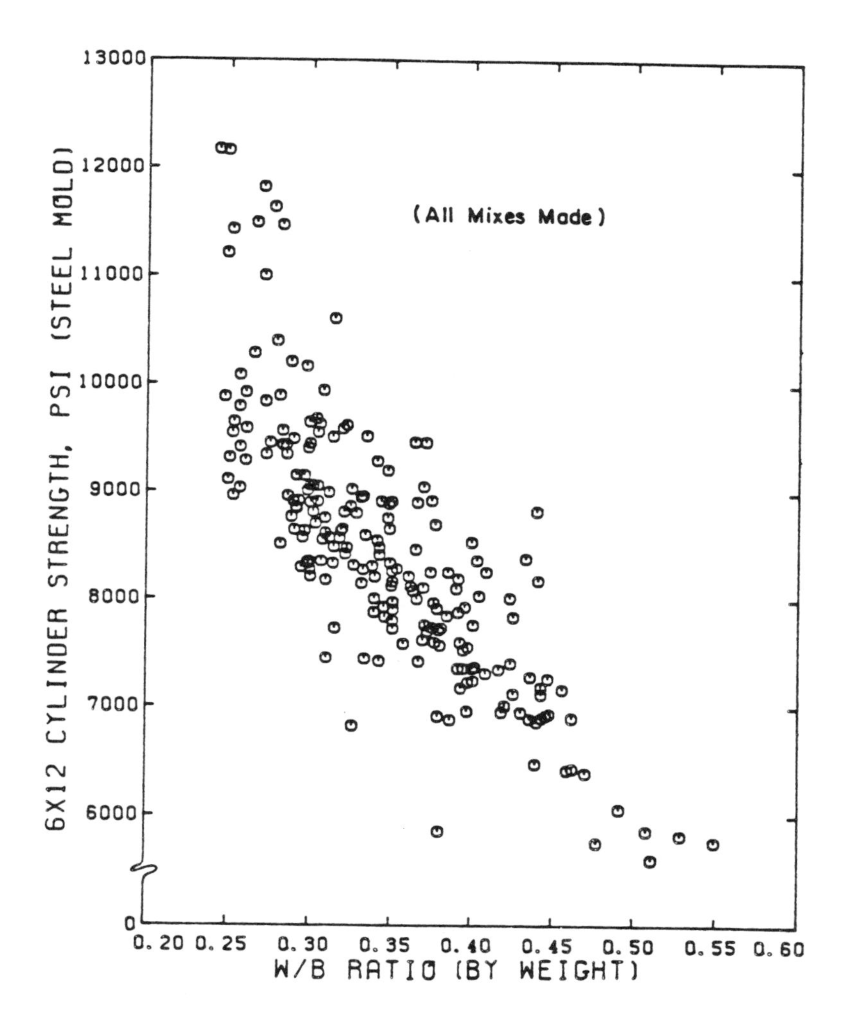

Fig. 5.3 Effect of water-binder ratio on the 28-day compressive strength of concrete for all 6-in. dia. x 12-in. cylinder specimens made, with and without chemical admixtures and fly ash.

5.4 Cement Type

For mixes containing no admixtures, it was found that type II cement was more suitable for production of high strength concrete than cement types I or III. However, for concretes having compressive strengths of about 8,500 psi or less, the effect of cement type and brand were inconclusive. Therefore, it is recommended that high strength concrete trial mix programs include a comparison of several available brands and types of cement, even though a given type or brand may have produced less desirable results when producing normal strength concrete.

For mixes made using superplasticizers, the highest compressive strengths were produced using type II cement. However, cement type has less of an effect on concrete compressive strength for mixes containing superplasticizers than on mixes containing no admixtures. In general, high strength concrete can be produced with any type of cement when a compatible superplasticizer is added to the mix.

Mixes made using type II cement had a lower mixing water requirement than mixes made with cement types I and II, with or without the addition of superplasticizer. An 8.5-sack mix containing superplasticizer and having a 4-in. slump could not be produced using a w/c ratio of 0.30 for cement types I and III. However, this was readily accomplished using type II cement. Higher compressive strength and denser concrete result from a lower mixing water content and lower heat of hydration at early ages such as that needed for workable mixes containing type II cement.

5.5 Superplasticizer Dose and Brand

Changing the brand of superplasticizer used in several concrete mixes resulted in no consistent change in the compressive strength of concrete. However, this is not expected to be a general rule. The effect of superplasticizer brand on compressive strength of concrete will depend greatly on the compatibility of the admixture with the other concrete-making materials. Each brand of superplasticizer must be tested individually for compatibility with the Portland cement and aggregates used, especially as it relates to workability, setting time, and compressive strength. For instance, the workability of concretes made with the two superplasticizer brands used in this study were noticeably different. Specifically, more bleeding occurred with brand A, while similar mixes made using admixture brand B were more cohesive and had better finishability. Both admixtures were satisfactory for producing high strength concrete.

Similarly, it is expected that the compatibility of different types of superplasticizer, such as naphthalene, melamine, and lignosulfates, with other concreting materials may affect compressive strength results.

In this study, superplasticizer dosage was found to be an important variable in the production of high compressive strength concrete. The required dosage for a given slump and w/b ratio depended on type and amount of cement, amount of fly ash, and mix proportions. For producing high strength concrete at a ready-mix plant in Texas, trial mixes will have to be performed not only to determine strength-producing properties of superplasticizers, but also to determine the dosage for optimum workability and placement characteristics. Superplasticizer dosages as recommended by the manufacturers of the admixtures used in this study were not adequate for the production of high strength except as a starting point for trial mixes. The effects of the types and amounts of cement and other materials used on the effectiveness of the superplasticizer are not necessarily accounted for in the manufacturer's recommended dosages. As shown in Fig. 5.4, increasing the superplasticizer dosage above manufacturer's recommendations resulted in an increase in concrete compressive strength if the dosage was not so high that it caused segregation of the mix.

In general, 7-sack mixes became unworkable at high superplasticizer dosages, especially when using a CA/FA ratio of 1.5 or higher. An upper limit of 15 fl.oz. of superplasticizer per 100 lbs of cement is recommended in 7.0-sack mixes for the mix proportions and materials used in this study. Compressive strengths at 28 days exceeded 9,000 psi for 7-sack mixes in some cases, but the workability was often poor and rocky and the specimens were very difficult to strike off and finish. Use of a CA/FA ratio of 1.0 or less, and use of 3/4-in. max. size coarse aggregate instead of 1/2-in. may be expected to improve the workability considerably in 7-sack mixes, but will likely reduce compressive strength.

For 8.5-sack mixes, superplasticizer dosages of 10 fl.oz./100 lbs of cement or more significantly increased the compressive strength

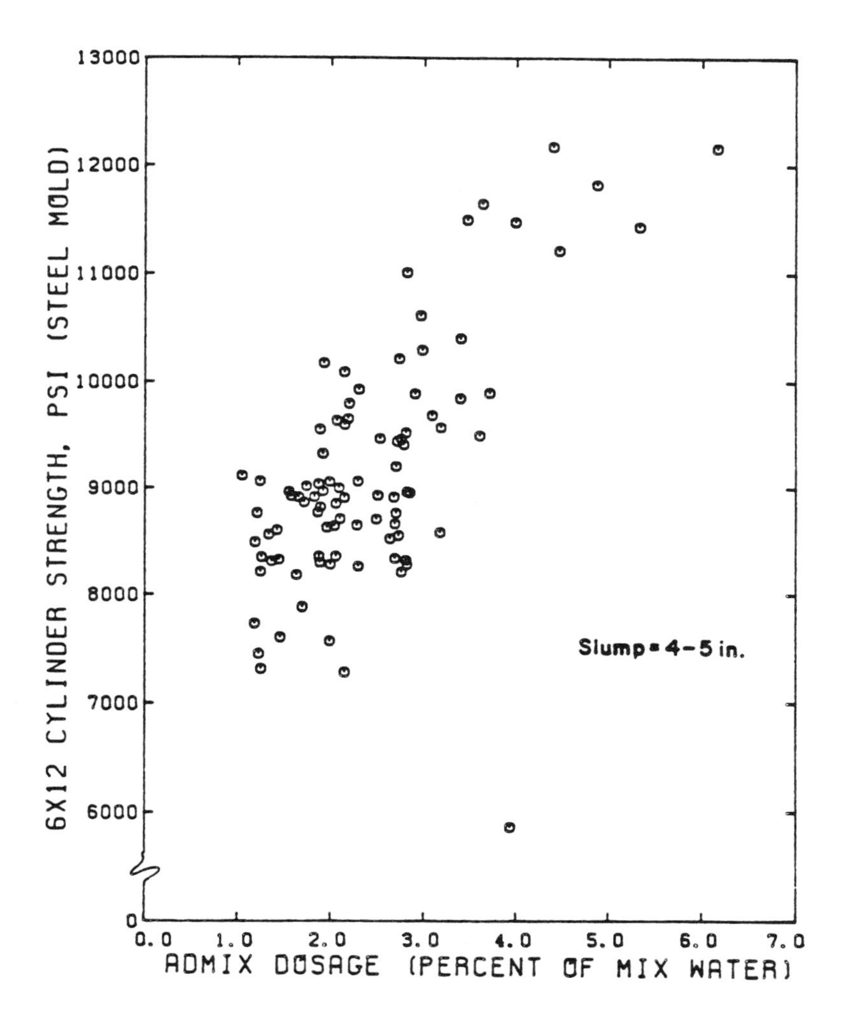

Fig. 5.4 Effect of superplasticizer dosage (expressed as a percent by weight of total mixing water) on the 28-day compressive strength of concrete for all mixes made containing superplasticizer.

at both 28 and 56 days. Workability of these mixes was generally very good. Finishability was much better than for 7.0-sack mixes.

Ten-sack mixes may also be expected to increase in strength for higher superplasticizer dosages, but not significantly, unless w/c ratios considerably below 0.30 are used. In general, the objective of using the superplasticizer was to obtain a w/c ratio close to 0.30. The reduction in w/c ratio for 8.5 sack mixes from about 0.37 to 0.30 resulted in higher concrete strengths. However, since the reduction in w/c ratio for 10-sack mixes was only from approximately 0.33 to 0.30, and since the total surface area of the cement particles was greater and therefore less affected by a given amount of superplasticizer than in an 8.5-sack mix, strength increases due to increases in dosage were less significant in 10-sack mixes.

5.6 Coarse Aggregate Size

A smaller coarse aggregate max. size is required for production of high strength concrete than for production of normal strength concrete when no chemical admixtures are used, as shown in Fig. 5.5. For a cement content of 7 sacks/cu.yd. the compressive strength of concrete was controlled by the w/c ratio in mixes containing no chemical or mineral admixtures. As a result, mixes made with 1-in. max. size coarse aggregate, which required the least mixing water for a given slump, produced the highest compressive strengths. For higher cement contents, however, using 1/2-in. max. size coarse aggregate resulted in the highest compressive strengths, despite the higher w/c ratio of these mixes.

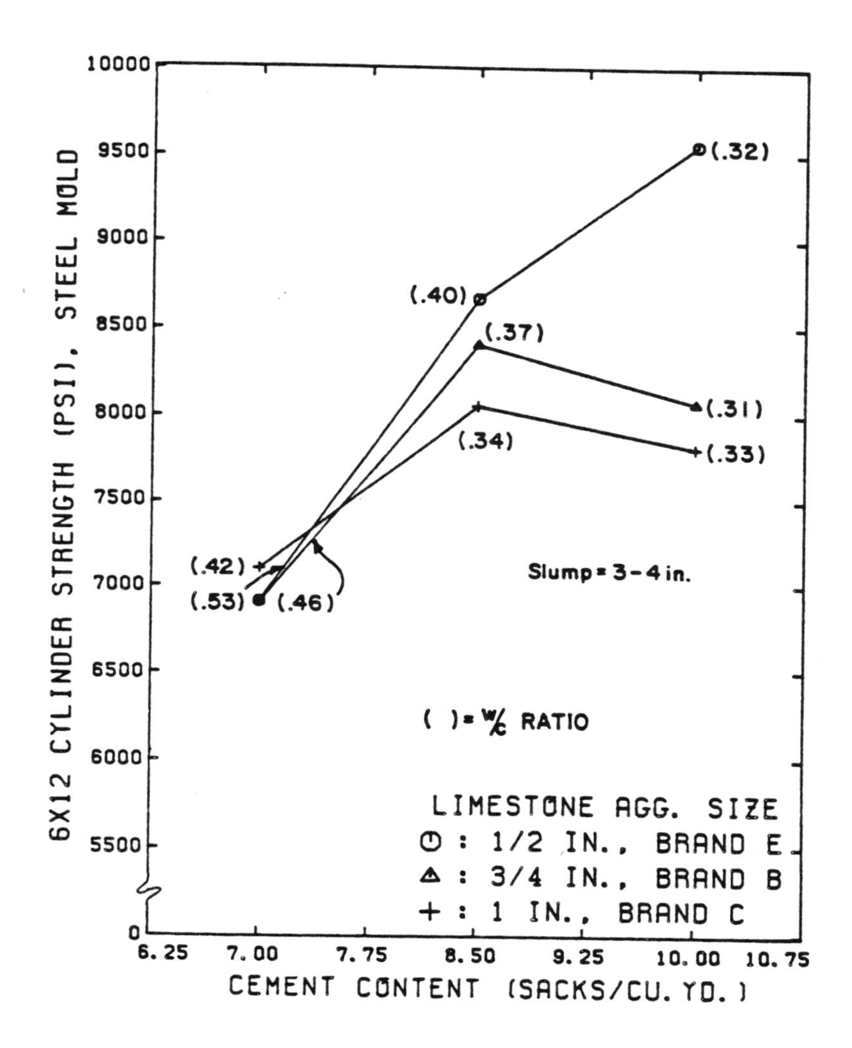

Fig. 5.5 Effect of coarse aggregate max. size and cement content on the 56-day compressive strength of concrete for mixes having a CA/FA ratio of 2.0 and made with type II cement, crushed limestone coarse aggregate, sand B, and no admixture.

The effect of increased homogeneity in concrete containing 1/2-in. coarse aggregate, higher cement contents and overall decreased average coarse aggregate-to-mortar bond stresses due to greater aggregate surface area apparently overcome the effect of the higher water requirement as the controlling factor for compressive strength. Disruptive stress concentrations at the aggregate-mortar interface may be less likely to occur until higher compressive stresses are applied with smaller size coarse aggregates. The strength of the mortar may come closer to being fully realized as well.

High strength concrete having a compressive strength of 9140 psi at 56 days was achieved using 1-in. max. size aggregate, 10 sacks of type II cement per cu.yd., a w/c ratio of 0.31, and no admixture. However, all 1/2-in. aggregate concrete mixes containing 10 sacks of type II cement per cu.yd. resulted in high compressive strengths ranging from 9000 psi to 9560 psi at 56 days with w/c ratios averaging 0.34.

High strength concrete may be produced using any maximum size of coarse aggregate ranging from 1/2-in. to 1-in. max. size when a superplasticizer is added. However, the greatest compressive strengths were achieved using 1/2-in. max. size coarse aggregate in this study.

Use of a CA/FA ratio of approximately 2.0 is also recommended for consistent production of high strength concrete.

The higher rate of strength gain between 28 and 56 days observed for higher w/c ratio mixes containing 1/2-in. aggregate is likely due to the extra available pore water which can enhance hydration at later ages.

5.7 Coarse Aggregate Gradation

Based on the limited information presented in Section 4.7, the effect of the gradation of the coarse aggregate on the compressive strength of high strength concrete is directly related to the effect of the gradation on the mixing water requirement for a given slump. As shown in that section, changes in gradation of the coarse aggregate resulted in a change in the w/c ratio of about 0.01. As a result, the difference in compressive strengths of the concretes made using the coarsest and finest gradations was nearly 1,000 psi, or 10 percent.

Due to the high fines content and the use of workability admixtures in the production of high strength concrete, the effect of variations in coarse aggregate gradation within allowable ASTM C-33 limits is not expected to be significant.

5.8 Coarse Aggregate Type

Based on the limited information from this study for producing concrete strengths greater than 9,500 psi, aggregate surface texture is very important. High strength concrete can be produced using gravel or limestone coarse aggregate without the need of adding superplasticizers to the concrete. However, mixes made using limestone aggregate generally resulted in higher concrete strengths, especially for high cement contents and 56 day test age. Improved bonding due to the rough surface of the aggregate and good mineralogical compatability between the limestone aggregate and the mortar are important in achieving very high concrete compressive strengths, especially in 10-sack mixes without admixtures and in 8.5 -sack mixes containing superplasticizers.

Dense limestone coarse aggregate having a DRUW of at least 90 to 92 lb/cu.ft., and a BSG_{ssd} of at least 2.55 is recommended for production of high strength concrete.

Figure 5.6 shows the failure surface of a 6-in. dia. x 12-in. high strength concrete specimen tested in compression. The smooth failure planes passed through the limestone aggregate, rather than around the aggregate. Gravel mixes resulted in similar compression test failure planes. However, in flexure beam tests, the failure surface always propagated through the limestone aggregate, while some aggregate bond failure often occurred in the failure plane of gravel mixes.

5.9 Sand Fineness

The higher the cement content, the smaller the difference in compressive strength of similar high strength concrete mixes produced using sands having fineness moduli of between 2.7 and 3.1. Based on the results presented in Chapter IV, it can be concluded that high strength concrete can be produced using sands whose fineness moduli are in the range from 2.7 to 3.1.

In concrete mixes containing superplasticizer, using sands with a fineness modulus as low as 2.4 resulted in high concrete compressive strengths. Contrary to the production of normal strength concretes, use of the finer sands generally resulted in higher strengths when producing high strength concretes. The finer sands allowed the use of larger superplasticizer dosages without inducing any workability problems in the concrete.

Fig. 5.6 Failure surface of high strength concrete compressive strength test specimen.

5.10 Fly Ash

It is highly recommended that trial mix programs for high strength concrete include the use of a Class C fly ash. As shown in Fig. 5.7, more compressive strength was gained by adding fly ash to a concrete mix than by adding an equal weight of Portland cement, for mixes having a ratio of the total fly ash weight to the combined weights of fly ash and Portland cement in the range of 20 to 30 percent. Substitution of fly ash by weight for 20 to 30 percent of the Portland cement in an ordinary concrete mix resulted in the production of high strength concrete with a substantially lower cement factor. Concrete strengths at 28 days of over 11,000 psi were achieved in mixes containing approximately 6 sacks of Portland cement per cu.yd. when both fly ash and superplasticizer were used in the mix.

The strength-producing properties of the calcium and silicon components of the fly ash apparently add substantially to concrete strength, especially since the mixing water demand of the fly ash is less than that of Portland cement.

The two sources of fly ash used in this study produced fly ash with very different total calcium and silicon contents. Yet their concrete strength-producing capacities were very similar.

It is recommended that as part of the trial mix design procedure, the engineer determine both the optimum total binder content and fly ash content for the materials available and given strength requirements.

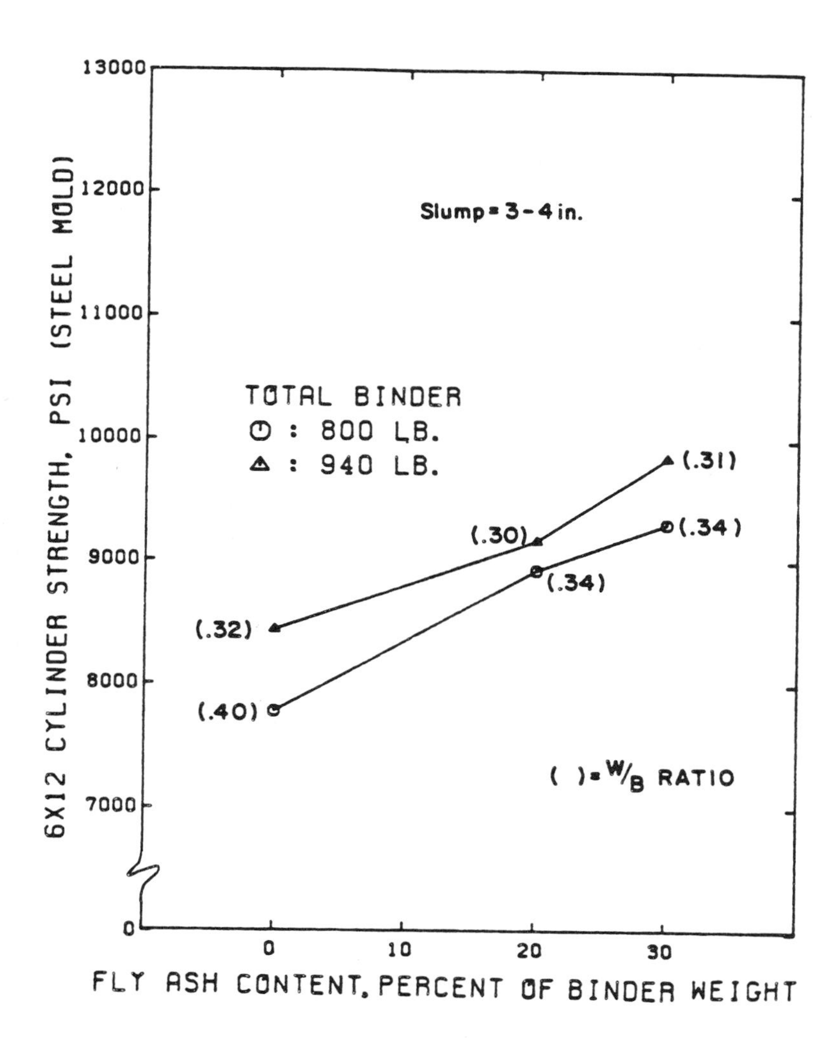

Fig. 5.7 Effect of total binder content and fly ash content on the 28-day compressive strength of concrete for mixes having a CA/FA ratio of 2.0 and made with type II cement, fly ash A, 1/2-in. limestone E, sand B, and no admixture.

Fly ash used in structural concrete should be laboratory-tested for compatibility with other materials before it is used in the field. It should also be tested at regular intervals during production for consistency of quality and composition. Changes in the operating procedures at the power plant boiler where the fly ash is collected can seriously affect the fly ash chemical composition.

5.11 Reducer-Retarders and Hot-Weather Concrete

The use of a reducer-retarder admixture in the low range of the manufacturers' recommended dosages makes it possible to produce high strength concretes even when mixed at temperatures above 100°F. High strength concrete can be produced even if retempering water has to be added to restore the workability of the concrete mix, if a reducer-retarder is used. Superplasticizer redosages, after 60 to 90 minutes of mixing at 100°F, improved the compressive strength of the concrete significantly. This was especially true in the presence of a reducer-retarder.

Careful control of admixture dosages is recommended when using both superplasticizers and reducing-retarding admixtures in a given mix. Their combined retarding action can cause the rate of hardening and 1-day compressive strength to be reduced significantly if the reducer-retarder dosage is not controlled properly.

5.12 High Strength Concrete and Curing Age

The effect of specimen age on concrete compressive strength is shown in Fig. 5.8. Substitution of fly ash for 30 percent of the cement in a concrete mix resulted in slightly lower 24-hour compressive

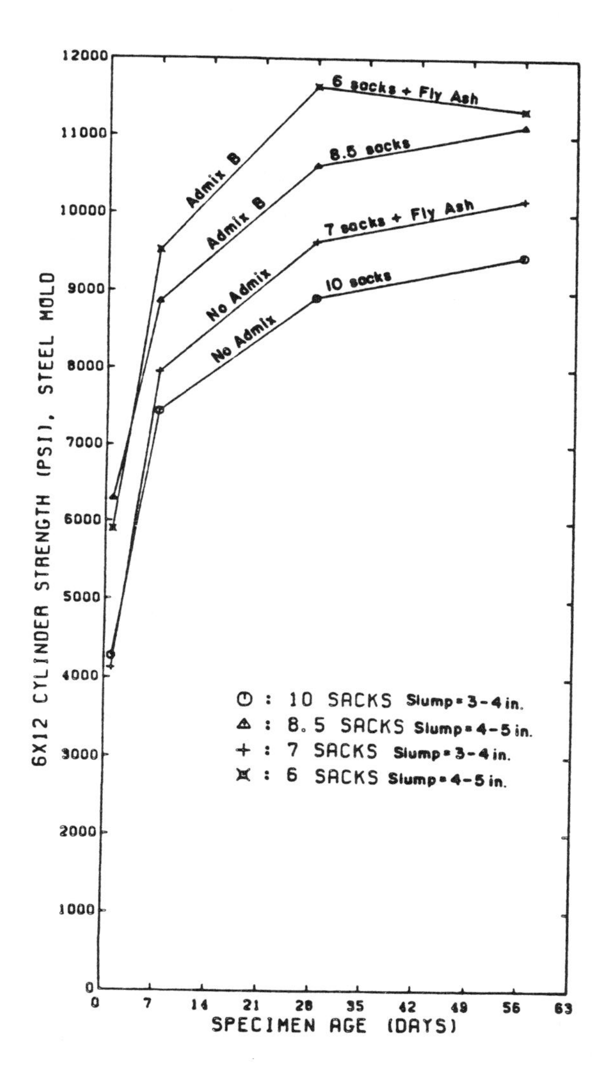

Fig. 5.8 Effect of specimen age and fly ash and a superplasticizer on the compressive strength of concrete for mixes having a CA/FA ratio of 2.0 and made with type II cement, fly ash A (0% or 30% by wt.), 1/2-in. limestone E, and sand B.

strength of concrete. However, superplasticizers increased the early strength of concrete with or without fly ash. Concrete with a superplasticizer dosage of between 15.0 and 23.0 fl.oz. per 100 lbs of Portland cement had a 24-hour compressive strength of 6,000 psi. Similar mixes containing no superplasticizer had 24-hour strengths of just over 4,000 psi.

Seven-day compressive strengths for these mixes ranged from about 7,500 psi to 9,500 psi. Compressive strengths at 28 days ranged from 11,000 to 12,000 psi were measured.

A reducer-retarder in a concrete mix can reduce the 24-hour strength by 20 percent or more, but the 7-day and later age compressive strengths were nearly the same as those of non-retarded mixes.

Construction which requires high 24-hour concrete strength may require a superplasticizer dosage of at least 15 oz/100 lbs of cement in an 8.5-sack mix. Superplasticizers may not be needed if 4,000 psi is an acceptable 24-hour compressive strength.

Type III cement may also improve the early strength of concrete, but it is expected to result in a slightly lower concrete strength at later ages.

5.13 Curing and Capping

Based on the limited information from this study, the compressive strength of high strength concrete is not reduced by drying until testing if the concrete has had at least 7 days of moist curing at a temperature of 70 to 80°F.

The type of capping compound used to cap concrete compression specimens affects the measured concrete compression strength. It is suggested that a high strength capping compound be used when testing high strength concrete.

5.14 Flexural Strength and Split Cylinder Strength

As shown in Fig. 5.9, the modulus of rupture for high strength concrete falls within the range from $8.0\sqrt{f'_c}$ and $12\sqrt{f'_c}$, regardless of mix proportions or materials used.

Split cylinder test results are of the order of 75 percent of the modulus of rupture.

5.15 Specimen Mold Size and Type

It is important to take into consideration the type and size of specimen mold used when evaluating compressive strength test results of high strength concrete. An increase in compressive strength of 10 percent can be expected when using 4-in. dia. x 8-in. cylinders instead of 6-in. x 12-in. cylinders, or when using same size cylinder molds made of steel rather than cardboard.

No conclusive results were obtained in limited comparisons of plastic molds with steel and cardboard molds.

5.16 Admixtures and Batching Procedures

It is recommended that cement be thoroughly moistened before superplasticizers and reducer-retarders are added to high strength concrete mixes. Hydration can be hindered greatly if dry cement particles are coated by superplasticizer before they are combined with

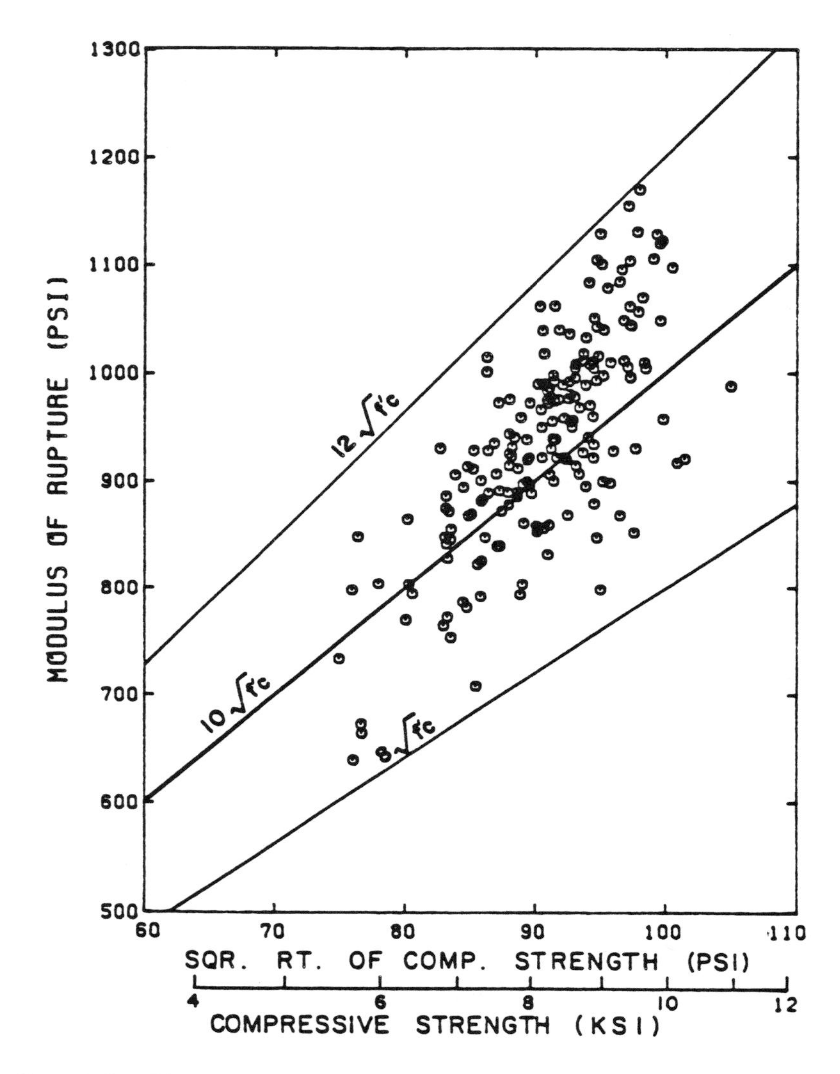

Fig. 5.9 Plot of modulus of rupture versus compressive strength for all flexure test beam specimens made in this study. Each point represents the average of three test results.

the water and as a consequence the quality of the fresh concrete could be adversely affected. In this study, addition of superplasticizer to unmoistened cement resulted in segregation and a slimy appearance of the fresh concrete. Addition of more water to the mix restored workability after sufficient mixing. However, control over the w/c ratio was lost, since more than the usual amount of mixing water was required for the desired slump.

If superplasticizers are used without other admixtures, at least half of the dose should be added to the concrete with the last portion of mixing water added. The remaining admixture should be added directly to the fresh concrete after mixing has started.

When superplasticizers and reducer-retarders are used together, the reducer-retarder should be added first with some of the mixing water, after the cement is moist. Then, after thorough mixing, the superplasticizer should be added as described above.

The two superplasticizers used in this project dispersed through the moistened fresh concrete very quickly after several revolutions of the mixer. However, special care must be taken to ensure thorough mixing and moistening of all materials in all parts of the concrete mixer when superplasticizers are added to high strength concrete mixes because of the low w/c ratio.

VI. Conclusions and Recommendations

6.1 Conclusions

The results of this study demonstrate that high strength concrete can be produced in the State of Texas with readily available materials using conventional batching procedures. The following conclusions have been made regarding the selection of materials, mix design, production, and testing of high strength concrete.

1. The water-cement, or water-binder, ratio is the most influential parameter affecting the compressive strength of high strength concrete. In general, to produce concrete having a 56-day compressive strength of at least 9,000 psi, the water-binder ratio must be less than 0.35.

2. A cement content of at least 10 sacks/cu.yd. is required to produce high strength concrete having a slump of 3 to 4 in., if no chemical or mineral admixture is added to the mix. A cement content of 8.5 sacks/cu.yd. is optimum for strength and workability of high strength concrete mixes containing superplasticizer, for a water-cement ratio of 0.30 and a slump of 4 to 5 in.

3. Compressive strength of concrete increases as superplasticizer dosage increases, up to a dosage which causes a concrete mix to become segregated and unworkable. The addition of too much superplasticizer to a high strength concrete mix may result in significant retardation of concrete hardening. The brand of superplasticizer used affects

both the workability and the compressive strength of high strength concrete.

4. High strength concrete can be produced using natural gravel or crushed stone. However, higher compressive strengths are obtained with concrete made using crushed stone.

5. For concrete mixes made with cement contents of 8.5 sacks/cu.yd. or more but without superplasticizers, using 1/2-in. max. size coarse aggregate results in higher concrete compressive strengths at 56 days for mixes having a similar slump. For concrete mixes made with a superplasticizer, use of any size of coarse aggregate between 1/2-in. and 1-in. can result in high compressive strength. However, the highest compressive strengths result from the use of 1/2-in. max. size coarse aggregate.

6. High strength concrete can be produced using a sand with a fineness modulus of from 2.7 to 3.1 for mixes containing no admixtures. Sands having a fineness modulus of as low as 2.4 are satisfactory for producing high strength concrete when a superplasticizer is used.

7. More compressive strength is gained by adding Class C fly ash to a concrete mix than by adding an equal weight of Portland cement, if the ratio of the weight of fly ash to the combined weights of fly ash and Portland cement is in the range from 20 to 30 percent.

8. The source of a fly ash affects the concrete strength-producing properties of the fly ash.

9. The 1-day strength of high strength concrete is slightly reduced by the addition of fly ash and can be significantly increased by the addition of superplasticizer.

10. The 28-day compressive strength of high strength concrete which has been cured under ideal conditions for 7 days after casting is not seriously affected by curing in hot and dry conditions from 7 to 28 days after casting.

11. The compressive strength of high strength concrete specimens cast using 4-in. dia. x 8-in. molds is 10 percent higher than that of concrete specimens cast using 6-in. dia. x 12-in. molds, in general. In general, the compressive strength of high strength concrete specimens cast in steel molds is 10 percent higher than that of concrete specimens cast in cardboard molds, in general.

12. The type of capping compounds used to cap high strength concrete compressive strength specimens for compression testing affects the test results. High strength capping compounds should be used.

13. The modulus of rupture of high strength concrete falls between $8.0\sqrt{f'_c}$ and $12\sqrt{f'_c}$.

14. High strength concrete having a slump of 4-in. or greater can be produced even when mixing temperatures are of the order of 100°F and the total period of mixing is between 60 min. and 90 min.

6.2 Mix Design Recommendations

Ten classes of regular and special concretes are presently specified in Tables 4 and 5 of the _1982 Standard Specifications for Construction of Highways, Streets, and Bridges_ of the Texas State

Department of Highways and Public Transportation. Using a format similar to that used by the Texas State Department of Highway and Public Transportation, Table 6.1, entitled "High Strength Concrete Mix Design Guidelines" is presented here. The information in Table 6.1 is a result of over 200 trial batches of concrete made using materials commercially available to ready-mix and precast plants in Texas and mixed using conventional mixing techniques. The recommendations are based on a study of the interaction among components of plain concrete and its mix proportions and of their contribution to the compressive strength of high strength concrete. It is expected that the recommendations presented in Table 6.1 will serve as a guideline to resident engineers in selection of materials and proportions for producing high strength concrete in the State of Texas. Table 6.1 is intended to be used as a guideline only, and it should not replace the making of trial mixes. As new information becomes available, the recommendations in Table 6.1 should be modified to incorporate field experience in using high strength concrete. Substantial improvements in strength and workability may be achieved simply by experimenting with different brands of cement, chemical admixtures and fly ash. Concrete producers are also encouraged to try larger coarse aggregates in concretes with superplasticizers, and fine aggregates with finer gradations. However, the aggregates and proportions described by Table 6.1 represent the optimum conditions for the materials studied. An increase in the amount of water used above that recommended may result in a drastic loss of compressive strength. Admixture dosages can be expected to vary with admixture brand. See the

TABLE 6.1 High Strength Concrete Mix Design Guidelines

Class (Reference Mix from this Study)	H-H-00 ("Q")	H-H-01 ("R")	H-H-10 ("S")	H-H-11 ("T")
Sacks cement per cu.yd.	10.0	8.5	7.0	6.0
Min. Comp. Str. (f'_c) 56 day, psi	9,500 (a)	10,500 (a)	10,000 (a)	11,000 (a)
Min. Beam Str. (f_r) 28 day, psi	1,040 (b)	1,130 (b)	960 (b)	990 (b)
Max Water-Cement Ratio (gal/sack)	3.9	3.4	5.0	4.5
Max Water-Binder Ratio (gal/100 lbs)	4.1	3.6	3.7	3.3
Crushed Coarse Aggr. No. (c)	Tx.Gr. 6 (c)	Tx.Gr. 6 (c)	Tx.Gr. 6 (c)	Tx.Gr. 6 (c)
CA/FA Ratio (by weight)	2.0 (d)	2.0 (d)	2.0 (d)	2.0 (d)
Fly Ash	---	---	Yes (g)	Yes (g)
Superplasticizer	---	Yes (e)	---	Yes (e)
General Usage				
Prestressed Concrete	Yes	Yes	Yes	Yes
Cast in Place	Yes	Yes	Yes	Yes
Other Notes				
Good Formed Surfaces	Yes	Yes	Yes	Yes
Good Finished Surfaces	Yes	See Note (f)	Yes	See Note (f)

TABLE 6.1 High Strength Concrete Mix Design Guidelines (continued)

Notes:

(a) Based on tests performed on 6 in. dia. x 12 in. cylinder of concrete made using a rigid steel mold.

(b) Based on tests performed on 6 in. x 6 in. x 18 in. simply supported beam with loads placed at third points.

(c) Crushed stone should have dry rodded unit weight of at least 90 lb/cu.ft., and a saturated-surface dry specific gravity of at least 2.55.

(d) Mixes containing no superplasticizer should be made using a coarse sand whose fineness modulus is at least 2.70.

(e) Dosage of superplasticizer should be highest possible without causing segregation or excessive retardation of fresh concrete.

(f) Smoothly finished surfaces possible with motor-driven finishing tools. Despite high fines content this mix is not easily finished by hand.

(g) Use of Class C fly ash at a rate of 30 percent by weight of the total cementitious material content is recommended for these mix proportions.

footnotes following Table 6.1 for additional important refinements to the guidelines.

The four classes of high strength concrete in Table 6.1 refer to mixes: (a) containing no fly ash or chemical admixtures; (b) containing a superplasticizer but no fly ash; (c) containing fly ash but no chemical admixture; and (d) containing a superplasticizer and fly ash.

6.3 Cost of High Strength Concrete per Cubic Yard

A compression member or prestressed girder made using high strength concrete can carry a greater load at a lower cost per cubic yard of concrete than if made using normal strength concrete. Based on material costs alone, the cost of a column made using 12,000 psi concrete is far less than the cost of a 6,000 psi concrete column designed to carry the same load with identical reinforcement.

Richart's study focussing on tied columns reported that the price per cubic yard of concrete increased by only 20 percent when the concrete compressive strength was increased from 3,000 psi to 6,000 psi [55]. However, use of the higher strength material resulted in overall savings of 25 percent. This savings included consideration of the replacement cost of a less durable concrete.

Based on typical material costs for the Austin area shown in Table 6.2, total costs per cubic yard for some concrete produced in this study are presented in Table 6.3.

It can be seen that the price per cubic yard of high strength concrete is more dependent on the relative quantities of cement, fly ash and admixtures used than on compressive strength. However, the cost of concrete per 1,000 psi decreases by over 20 percent with an increase in concrete compressive strength from 9,500 psi to 12,000 psi. Based on material costs and load-carrying capacity alone, it is most economical to use 12,000 psi concrete containing 6 sacks of cement per cubic yard, fly ash, superplasticizer, and a water reducer-retarder.

TABLE 6.2 Assumed Material Costs
(based on 1983 Prices
Austin, Texas)

Cement	$ 3.00/sack
Coarse Aggregate	$ 5.25/ton
Fine Aggregate	$ 3.50/ton
Fly Ash	$30.00/ton
Superplasticizer	$ 5.25/gal (4.1¢/fl.oz.)
Reducer-retarder	$ 5.00/gal (3.9¢/fl.oz.)

TABLE 6.3 Comparison of Costs of High Strength Concretes

Concrete Description	Approximate Mix Design		Total Cost Material $/cu.yd.	Concrete per Cu.Yd. %/cu.yd.	Total Cost $/1000 psi	Rel. Cost
6.3 sack, 5,000 psi mix,	Cement	590 lb	$18.80	75%		
no admixtures	Coarse	1450 lb	3.81	15	$5.02	1.00
(for cost comparison only)	Fine	1430 lb	2.50	10		
			$25.51	100%		
10-sack, 9,500 psi mix,	Cement	940 lb	$30.00	83%		
no admixtures	Coarse	1830 lb	4.80	13	$3.82	0.76
(Mixes "Q")	Fine	870 lb	1.52	4		
			$36.32	100%		
8.5-sack, 10,500 psi mix,	Cement	800 lb	$25.50	68%		
with superplasticizer,	Coarse	2040 lb	5.36	14		
no fly ash,	Fine	1010 lb	1.77	5	$3.82	0.76
(Mixes "R")	Super.	15 fl.oz./100	4.92	13		
			$37.44	100%		
7.0-sack, 10,000 psi mix,	Cement	660 lb	$21.00	67%		
with fly ash,	Coarse	1820 lb	4.78	15	$3.58	0.71
no superplasticizer,	Fine	920 lb	1.61	5		
(Mixes "S")	Fly Ash	280 lb	4.20	13		
			$31.59	100%		

(continued)

TABLE 6.3 Comparison of Costs of High Strength Concretes (continued)

Concrete Description	Approximate Mix Design		Total Cost Material $/cu.yd.	Concrete per Cu.Yd. %/cu.yd.	Total Cost $/1000 psi	Rel. Cost
6.0-sack, 11,000 psi mix,	Cement	560 lb	$18.00	54%		
with fly ash	Coarse	2040 lb	5.36	16		
and superplasticizer	Fine	1040 lb	1.82	6	$2.99	0.60
(Mixes "T")	Fly Ash	240 lb	3.60	11		
	Super.	18 fl.oz./100	4.13	13		
			$32.91	100%		
6.0-sack, 12,000 psi mix,	Cement	560 lb	$18.00	51%		
with fly ash,	Coarse	2080 lb	5.46	15		
superplasticizer	Fine	1040 lb	1.82	5	$2.98	0.60
and reducer-retarder	Fly ash	240 lb	3.60	10		
	Super.	25 fl.oz./100	5.74	16		
	Red.-ret.	5 fl.oz./100	1.09	3		
			$35.71	100%		

Appendix A: Material Properties

The physical properties of the materials used in the study reported herein, including aggregates, cements, and fly ashes, are presented in this section.

Each material used for this study is designated by a capital letter, indicating a source or brand, followed by a number referring to the delivery date of the material. This designation is used throughout this report.

After coarse aggregate A was used for a small number of mixes, this material became unavailable. Since the mix series using aggregate A could not be completed, the data for the concrete mixes made using aggregate A was not used in this report. Coarse aggregate A is excluded from this appendix.

It was determined that fine aggregate A did not meet ASTM requirements for fine aggregates used in structural concrete. The few mixes made using it were not discussed in this report. Fine aggregate A is excluded from this appendix also.

TABLE A.1 Chemical and Physical Properties of Cements Used in This Study

Type / Identification	I A1	I A2	I-II B1	II C1	II C2	II C3	II C4	II C5	II C6	II C7	I D1	I D2	III E1	III E2
SiO_2 %	----	----	22.88	22.02	21.38	22.02	21.24	21.24	22.70	22.68	20.0	20.2	20.00	20.00
Al_2O_3 %	5.2	5.2	3.83	4.91	4.79	4.38	5.11	5.11	3.82	4.04	5.8	6.0	5.7	5.8
Fe_2O_3 %	3.3	3.1	3.75	4.83	4.69	4.88	4.71	4.71	3.88	3.76	2.6	2.9	2.6	2.9
CaO %	----	----	65.31	64.18	64.47	64.24	64.27	64.27	65.81	65.30	65.4	65.3	65.3	64.8
MgO %	1.0	1.0	0.95	0.80	0.80	0.79	0.87	0.87	0.80	0.83	0.80	0.89	0.86	0.92
SO_3 %	2.7	2.7	2.09	2.52	2.73	2.60	2.33	2.33	2.37	2.33	2.9	2.9	3.25	4.2
Ignition Loss %	1.1	1.0	0.87	0.56	0.66	0.66	0.53	0.53	0.55	0.42	----	----	----	----
Na_2O Equiv. %	0.46	0.50	0.45	0.57	0.55	0.41	0.47	0.47	0.47	0.51	0.55	0.60	0.49	0.60
C_3S %	----	----	54.91	48.66	53.29	50.35	52.50	52.50	57.42	54.30	63.3	59.6	62.5	56.7
C_2S %	----	----	24.19	25.73	21.10	25.15	21.29	21.29	21.76	24.06	9.6	12.9	10.1	14.5
C_3A %	8.0	8.4	3.80	4.84	4.75	3.35	5.57	5.57	3.56	4.35	11.0	11.0	10.7	10.5
C_4AF %	10.04	9.43	11.42	14.69	14.27	14.85	14.33	14.33	11.81	11.44	7.9	8.8	7.9	8.8
Fines (Wagner)	2000	1900	1200	1942	1942	1938	1985	1985	1974	1909	1905	1870	2870	2640
(Blaine)	3559	3559	2240	3383	3383	3617	3536	3536	3336	3242	3755	3765	5895	5600
Set: Vicat (Gilmore)	1:50	2:50	----	(2:35)	(2:35)	(2:28)	(2:20)	(2:20)	(2:25)	(2:30)	3:00	2:40	1:50	2:45
max/min	3:20	4:10	----	(5:05)	(5:05)	(4:52)	(4:45)	(4:45)	(5:00)	(5:10)	5:00	4:55	3:50	4:10
False Set	88.0	----	----	84.4	84.4	75.5	89.5	89.5	85.2	93.3	----	----	----	----
Autoclave	.030	.032	.018	.008	.008	.012	.011	.011	.011	.006	----	----	----	----
Air %	9.3	9.0	----	9.4	9.4	8.4	9.4	9.4	9.6	9.2	----	----	----	----
Ins. Res.	0.19	0.12	----	0.18	0.18	0.17	0.15	0.15	0.11	0.15	----	----	----	----
1-day, psi	----	----	----	1383	1383	1515	1649	1649	1315	1315	----	----	3400	3605
3-day, psi	3483	3342	----	2671	2671	2813	3048	3048	2662	2683	3635	3505	5150	4800
7-day, psi	4483	4725	4616	3846	3846	3752	4075	4075	3630	3663	4360	4400	5925	5790
28-day, psi	----	----	5866	5800	5800	5874	5975	5975	5373	5646	6465	6050	7040	7300

TABLE A.2 Chemical and Physical Properties of Fly Ash Used in this Study.

	Fly Ash A	Fly Ash B
SiO_2 (% wt.)	35.96	63.1
Al_2O_3 (% wt.)	19.81	12.9
Fe_2O_3 (% wt.)	5.02	5.25
CaO (% wt.)	27.24	11.2
MgO (% wt.)	4.91	2.52
SO_3 (% wt.)	3.15	1.46
Na_2O (% wt.)	2.23	0.34
K_2O (% wt.)	0.42	0.47
Loss on Ignition (% wt.)	0.41	0.45
Moisture (% wt.)	0.02	0.06
Ammonia (% wt.)	---	0.031
% Retained on No. 325 Sieve	15.0	15.8
Pozzolanic Activity at 28 Days (% of Control)	87.3	95.8
Water Requirement (% of Control)	89.6	89.2
Specific Gravity	2.62	2.50
Autoclave Soundness	.104	---

Applicable Specifications for Aggregate Tests [54, 66, 67]

- Item 421, Concrete for Structures, 1982 Standard Specifications for Construction of Highways, Streets and Bridges, Texas State Department of Highways and Public Transportation.

- ASTM C33-80, Standard Specification for Concrete Aggregates.

- ASTM C29-78, Standard Test Method for Unit Weight and Voids in Aggregate; Tex-404-A, Determination for Unit Weight of Aggregate.

- ASTM C136-80, Standard Method for Sieve Analysis of Fine and Coarse Aggregates; Tex-401-A, Sieve Analysis of Fine and Coarse Aggregates.

- ASTM C127-80, Standard Test Method for Specific Gravity and Absorption of Coarse Aggregate; ASTM C128-79, Standard Test Method for Specific Gravity and Absorption of Fine Aggregate; Tex-403-A, Saturated Surface-Dry Specific Gravity and Absorption of Aggregates.

- ASTM C566-78, Standard Test Method for Total Moisture Content of Aggregate by Drying; Tex-409-A, Free Moisture in Aggregates for Concrete.

TABLE A.3 Coarse Aggregate B1

Material: crushed limestone (yellow-white in color)
Max. Size: 3/4 in.

Bulk specific gravity, SSD: 2.59 (ASTM C127)
Apparent specific gravity: 2.70
Absorption: 2.6%

Dry rodded unit weight: 95 lb/cu.ft. (ASTM C29)

Sieve Size	% Passing	ASTM C-33 Size 67	Texas Item 421 Gr. 5
1"	100	100	(same as ASTM)
3/4"	97.7	90-100	
1/2"	76.9	----	
3/8"	55.3	20-55	
1/4"	23.0	----	
4	8.6	0-10	
Pan	0	----	

TABLE A.4 Coarse Aggregate B2

Material: crushed limestone (yellow-white in color)
Max. Size: 3/4 in.

Bulk specific gravity, SSD: 2.63 (ASTM C127)
Apparent specific gravity: 2.71
Absorption: 1.75%

Dry rodded unit weight: 96 lb/cu.ft. (ASTM C29)

Sieve Size	% Passing	ASTM C-33 Size 67	Texas Item 421 Gr. 5
1"	100	100	(same as ASTM)
3/4"	94.8	90-100	
1/2"	61.8	----	
3/8"	37.2	20-55	
4	3.0	0-10	
10	0.8	----	
Pan	0	0	

TABLE A.5 Coarse Aggregate C1

Material: crushed limestone (yellow-white in color)
Max. Size: 1 in.

Bulk specific gravity, SSD: 2.57 (ASTM C127)
Apparent specific gravity: 2.70
Absorption: 3.2%

Dry rodded unit weight: 99 lb/cu.ft. (ASTM C29)

Sieve Size	% Passing	ASTM C-33 Size 57	Texas Item 421 Gr. 4
1-1/2"	100	100	(same as ASTM)
1"	91.7	95-100	
3/4"	75.1	----	
1/2"	57.4	25-60	
3/8"	43.6	----	
1/4"	22.3	----	
4	9.8	0-10	
Pan	0	0	

TABLE A.6 Coarse Aggregate D1

Material: crushed limestone (white in color)
Max. Size: 1/2 in.

Bulk specific gravity, SSD: 2.46 (ASTM C127)
Apparent specific gravity: 2.62
Absorption: 4.2%

Dry rodded unit weight: 85 lb/cu.ft. (ASTM C29)

Sieve Size	% Passing	ASTM C-33 Size 8	Texas Item 421 Gr. 7
1/2"	100	100	100
3/8"	86.5	85-100	70-95
1/4"	20.1	----	----
4	7.1	10-30	0-25
8	2.6	0-10	----
Pan	0	0	0

TABLE A.7 Coarse Aggregate E1

Material: crushed limestone (gray in color)
Max. Size: 1/2 in.

Bulk specific gravity, SSD: 2.65 (ASTM C127)
Apparent specific gravity: 2.74
Absorption: 1.9%

Dry rodded unit weight: 97 lb/cu.ft. (ASTM C29)

Sieve Size	% Passing	ASTM C-33 Size 7	Texas Item 421 Gr. 6
3/4"	100	100	(same as ASTM)
1/2"	99.0	90-100	
3/8"	79.0	40-70	
1/4"	25.1	----	
4	9.4	0-15	
8	2.3	0-5	
Pan	0	0	

TABLE A.8 Coarse Aggregate E2

Material: crushed limestone (gray in color)
Max. Size: 1/2 in.

Bulk specific gravity, SSD: 2.64 (ASTM C127)
Apparent specific gravity: 2.74
Absorption: 2.1%

Dry rodded unit weight: 95 lb/cu.ft. (ASTM C29)

Sieve Size	% Passing	ASTM C-33 Size 7	Texas Item 421 Gr. 6
3/4"	100	100	(Same as ASTM)
1/2"	99.5	90-100	
3/8"	80.2	40-70	
1/4"	25.8	----	
4	12.6	0-15	
8	2.6	0-5	
Pan	0	0	

TABLE A.9 Coarse Aggregate E3

Material: crushed limestone (gray in color)
Max. Size: 1/2 in.

Bulk specific gravity, SSD: 2.64 (ASTM C127)
Apparent specific gravity: 2.72
Absorption: 1.9%

Dry rodded unit weight: 93 lb/cu.ft. (ASTM C29)

Sieve Size	% Passing	ASTM C-33 Size 7	Texas Item 421 Gr. 6
3/4"	100	100	(same as ASTM)
1/2"	99.8	90-100	
3/8"	75.3	40-70	
4	5.1	0-15	
10	1.6	----	
Pan	0	0	

TABLE A.10 Coarse Aggregate E4

Material: crushed limestone (gray in color)
Max. Size: 1/2 in.

Bulk specific gravity, SSD: 2.68 (ASTM C127)
Apparent specific gravity: 2.74
Absorption: 1.2%

Dry rodded unit weight: 95 lb/cu.ft. (ASTM C29)

Sieve Size	% Passing	ASTM C-33 Size 7	Texas Item 421 Gr. 6
3/4"	100	100	(same as ASTM)
1/2"	99.5	90-100	
3/8"	66.1	40-70	
4	2.7	0-15	
8	0.5	0-5	
Pan	0	0	

TABLE A.11 Coarse Aggregate F1

Material: river gravel
Max. Size: 1/2 in.

Bulk specific gravity, SSD: 2.58 (ASTM C127)
Apparent specific gravity: 2.64
Absorption: 1.5%

Dry rodded unit weight: 97 lb/cu.ft. (ASTM C29)

Sieve Size	% Passing	ASTM C-33 Size 8	Texas Item 421 Gr. 7
3/4"	100	100	100
1/2"	99.8	100	100
3/8"	97.0	85-100	70-95
1/4"	49.1	---	---
4	7.2	10-30	0-25
8	0.5	0-10	0
Pan	0	0	0

TABLE A.12 Coarse Aggregate F2

Material: river gravel
Max. Size: 1/2 in.

Bulk specific gravity, SSD: 2.58 (ASTM C127)
Apparent specific gravity: 2.62
Absorption: 0.8

Dry rodded unit weight: 96 (ASTM C29)

Sieve Size	% Passing	ASTM C-33 Size 7	Texas Item 421 Gr. 6
3/4"	100	100	(same as ASTM)
1/2"	91.5	90-100	
3/8"	44.6	40-70	
4	1.9	0-15	
8	0.2	0-5	
Pan	0	0	

TABLE A.13 Fine Aggregate B1

Material: natural sand
Fineness modulus: 3.08

Bulk specific gravity, SSD: 2.56 (ASTM C128)
Apparent specific gravity: 2.60
Absorption: 1.0%

Dry rodded unit weight: 102 lb/cu.ft. (ASTM C29)

Sieve Size	% Passing	ASTM C-33	Texas Item 421 Gr. 1
3/8"	100	100	100
4	97.5	95-100	95-100
8	83.7	80-100	80-100
16	65.2	50-85	50-85
30	35.6	25-60	25-65
50	7.6	10-30	10-35
100	2.1	2-10	0-10
200	----	----	0-3
Pan	0	0	0

TABLE A.14 Fine Aggregate B2

Material: natural sand
Fineness modulus: 2.57

Bulk specific gravity, SSD: 2.57 (ASTM C128)
Apparent specific gravity: 2.64
Absorption: 1.8%

Dry rodded unit weight: 105 lb/cu.ft. (ASTM C29)

Sieve Size	% Passing	ASTM C-33	Texas Item 421 Gr. 1
3/8"	100	100	100
4	96.7	95-100	95-100
8	86.7	80-100	80-100
16	76.4	50-85	50-85
30	53.7	25-60	25-65
50	23.5	10-30	10-35
100	4.5	2-10	0-10
200	---	---	0-3
Pan	0	0	0

TABLE A.15 Fine Aggregate B3

Material: natural sand
Fineness modulus: 2.85

Bulk specific gravity, SSD: 2.57 (ASTM C128)
Apparent specific gravity: 2.64
Absorption: 1.5%

Dry rodded unit weight: 107 lb/cu.ft. (ASTM C29)

Sieve Size	% Passing	ASTM C-33	Texas Item 421 Gr. 1
3/8"	100	100	100
4	100	95-100	95-100
8	89.5	80-100	80-100
16	69.6	50-85	50-85
30	40.3	25-60	25-65
50	13.2	10-30	10-35
100	2.4	2-10	0-10
200	----	----	0-3
Pan	0	0	0

TABLE A.16 Fine Aggregate B4

Material: natural sand
Fineness modulus: 2.77

Bulk specific gravity, SSD: 2.56 (ASTM C128)
Apparent specific gravity: 2.63
Absorption: 1.7%

Dry rodded unit weight: 103 lb/cu.ft. (ASTM C29)

Sieve Size	% Passing	ASTM C-33	Texas Item 421 Gr. 1
3/8"	100	100	100
4	99.4	95-100	95-100
8	87.4	80-100	80-100
16	69.9	50-85	50-85
30	47.4	25-60	25-65
50	15.8	10-30	10-35
100	2.9	2-10	0-10
200	----	----	0-3
Pan	0	0	0

TABLE A.17 Fine Aggregate C1

Material: natural sand
Fineness modulus: 2.72

Bulk specific gravity, SSD: 2.62 (ASTM C128)
Apparent specific gravity: 2.69
Absorption: 1.6%

Dry rodded unit weight: 108 lb/cu.ft. (ASTM C29)

Sieve Size	% Passing	ASTM C-33	Texas Item 421 Gr. 1
3/8"	100	100	100
4	98	95-100	95-100
8	85.2	80-100	80-100
16	71.5	50-85	50-85
30	57.1	25-60	25-65
50	14.1	10-30	10-35
100	2.3	2-10	0-10
200	----	----	0-3
Pan	0	0	0

TABLE A.18 Coarse Aggregate C2

Material: natural sand
Fineness modulus: 2.45

Bulk specific gravity, SSD: 2.64 (ASTM C128)
Apparent specific gravity: 2.70
Absorption: 1.4%

Dry rodded unit weight: 104 lb/cu.ft. (ASTM C29)

Sieve Size	% Passing	ASTM C-33	Texas Item 421 Gr. 1
3/8"	100	100	100
4	98.3	95-100	95-100
8	87.7	80-100	80-100
16	77.7	50-85	50-85
30	66.1	25-60	25-65
50	22.5	10-30	10-35
100	2.1	2-10	0-10
200	----	----	0-3
Pan	0	0	0

TABLE A.19 Fine Aggregate D1

Material: natural sand
Fineness modulus: 2.75

Bulk specific gravity, SSD: 2.62 (ASTM C128)
Apparent specific gravity: 2.66
Absorption: 1.0%

Dry rodded unit weight: 106 lb/cu.ft. (ASTM C29)

Sieve Size	% Passing	ASTM C-33	Texas Item 421 Gr. 1
3/8"	100	100	100
4	94.7	95-100	95-100
8	81.6	80-100	80-100
16	73.7	50-85	50-85
30	55.7	25-60	25-65
50	17.5	10-30	10-35
100	2.2	2-10	0-10
200	----	----	0-3
Pan	0	0	0

Appendix B: Mixing and Testing Data

In the following pages, Table B.1 presents the test results and a list of all materials and proportions used for the concrete mixes made during the experimental phase of this study. In the test results for each mix, the numbers marked with asterisks are the averages and are the values used throughout Chapter IV. Every mix is identified by the five (four if no fly ash added) letters listed under "Mix I.D.". They stand for the cement brand, the coarse aggregate source, the fine aggregate source, the fly ash brand, and the chemical admixture brand, respectively. Some mixes were produced more than once and are marked with circled letters for easier comparison with the information presented in Chapters IV, V, and VI.

<u>Key</u>

(In order of appearance on table headings)

MIX I.D.	- Refers to cement brand, coarse aggregate brand, fine aggregate brand, fly ash brand (blank means none), and chemical admixture brand ("0" means none), respectively.
MIX DATE	- Month/day/year.
CF	- Cement factor (sacks/cu.yd.).
CAFA	- Coarse aggregate to fine aggregate weight ratio.
BRAND	- Identified by letter and number, as described in Appendix A.
TYPE	- Cement type (I, II, or III).

LBS/CUYD	- Total weight of the material per cu.yd. of concrete.
PCT VOLUME	- Total volume of the material as a percentage of the total volume of concrete. (Note: air content assumed at 2 percent.)
PCT REPLACED	- The weight of cement replaced by an equal weight of fly ash, expressed as a percentage of the total weight of fly ash plus portland cement.
SIZE	- Coarse aggregate maximum size (inches).
MATERIAL	- Type of coarse aggregate.
SOURCE	- See BRAND.
P:VOL	- See PCT VOLUME.
P:DRUW	- Total amount of coarse aggregate used per unit volume, expressed as a percentage of dry rodded unit weight.
FINENESS	- Fineness modulus of the sand.
CA/FA	- See CAFA.
TYPE	- Admixture type (superplasticizer and/or water/reducer retarder).
DOSE (OZ/100)	- Dosage of admixture in fluid ounces per 100 lbs of Portland cement.
(2ND TYPE)	- Same as TYPE. Assumes the use of two chemical admixtures.
(2ND DOSE)	- Same as DOSE (OZ/100). Assumes the use of two chemical admixtures.
W/C	- Water-cement ratio by weight.
W/B	- Water-binder ratio by weight.
GAL/SACK	- Water-cement ratio expressed as gallons of water per sack of portland cement.
PCT AIR	- Air volume (assumed to be 2 percent).
SLUMP	- Slump (inches).

UNIT WT	- Unit weight of fresh concrete (lb/cu.ft.).
MIX TMP	- Temperature of the concrete when it is cast (^{o}F).
MX TIME	- Period of time concrete is mixed (min.).
CURING	- Method of curing. Generally in moisture room at 73°F.
6x12, 6x6x18, 4x8	- Size of molds in inches.
STEEL	- Mold material.
CARDBD	- Mold material (cardboard).
PSI	- Pounds per square inch.
28-DAY	- Test age is 28 days after casting.

Additional Explanations

"COMPR STEEL 7D"	- Means compression specimen cast in steel mold and tested 7 days after casting. (PLSTC means plastic mold. CRDBD means cardboard mold.)
"SPLIT CRDBD 28D"	- Means split cylinder test specimen cast in cardboard mold and tested 28 days after casting.
"VIBRTD"	- Vibrated.
"MOIST", "DRY", "HOT & DRY", "UNDER WATER"	- All refer to types of curing and are followed by the number of days cured under that condition.
"HIGH STRNGTH CAP"	- High strength capping compound.
"MISC CAP"	- Another capping compound.

TABLE B.1

Mixing and Testing Data for Concrete Mixes
Made in This Study

	CEMENT	FLYASH	COARSE AGG	FINE AGG	ADMIXTURE	WATER	MISC	TEST RESULTS				
MIX I.D. MIX DATE CF/CAFA	BRAND TYPE LBS/CUYD PCT VOLUME	BRAND CLASS LBS/CUYD PCT VOLUME PCT REPLACED	SIZE MATERIAL SOURCE LBS/CUYD P:VOL,DRUM	FINENESS SOURCE LBS/CUYD PCT VOLUME CA/FA(LB/LB)	TYPE BRAND DOSE(OZ/100) (2ND TYPE) (2ND DOSE)	W/C W/B LB/CUYD GAL/SACK PCT AIR	SLUMP UNIT WT MIX TMP MX TIME CURING	6 X 12 CYLINDER (STEEL) (PSI)	6 X 12 CYLINDER (STEEL) (PSI)	6X6X18 BEAM (STEEL) (PSI)	4 X 8 CYLINDER (CARDBD) (PSI)	4 X 8 CYLINDER (STEEL) (PSI)
ABB 0	BRAND A2		3/4	3.10		.395	4.00	(28-DAY)	(57-DAY)	(28-DAY)	(28-DAY)	(28-DAY)
2/22/82	I	NONE	LIMESTONE	BRAND B1	NONE	.395	149.	• 7530. •	7750. •	936. •	7400. •	-0•
8.5/1.0	811.	0	BRAND B1	1365.	0	320.	76.0	7460.	7340.	958.	7280.	-0
	15.	0	1399.	32.		4.4	15. MIN	7940.	8980.	908.	7680.	-0
		0	32. 55.	1.02	-0	2.0	DAMP,73F	7200.	7820.	942.	7240.	-0
ABB 0	BRAND A2		3/4	3.10		.392	4.00	(28-DAY)	(57-DAY)	(28-DAY)	(28-DAY)	(28-DAY)
2/22/82	I	NONE	LIMESTONE	BRAND B1	NONE	.392	149.	• 7600. •	7620. •	892. •	7570. •	-0•
8.5/1.5	810.	0	BRAND B1	1097.	0	318.	76.0	7520.	7480.	875.	7760.	-0
	15.	0	1677.	25.		4.4	15. MIN	7820.	8190.	908.	7440.	-0
		0	38. 65.	1.53	-0	2.0	DAMP,73F	7460.	7200.	0	7520.	-0
ABB 0	BRAND A2		3/4	3.10		.376	3.50	(28-DAY)	(56-DAY)	(28-DAY)	(28-DAY)	(28-DAY)
2/23/82	I	NONE	LIMESTONE	BRAND B1	NONE	.376	151.	• 7970. •	7970. •	939. •	8650. •	-0•
8.5/2.1	816.	0	BRAND B1	909.	0	307.	76.0	8210.	7300.	958.	8870.	-0
	15.	0	1890.	21.		4.2	15. MIN	7480.	8720.	950.	8590.	-0
		0	43. 74.	2.08	-0	2.0	DAMP,73F	8210.	7590.	908.	8480.	-0
BBB 0	BRAND B1		3/4	3.10		.417	3.50	(28-DAY)	(56-DAY)	(28-DAY)	(28-DAY)	(28-DAY)
2/23/82	I	NONE	LIMESTONE	BRAND B1	NONE	.417	149.	• 7350. •	7760. •	792. •	6580. •	-0•
8.5/1.0	802.	0	BRAND B1	1340.	0	334.	76.0	7500.	7570.	792.	6050.	-0
	15.	0	1335.	31.		4.7	15. MIN	7670.	7660.	842.	6640.	-0
		0	32. 54.	1.04	-0	2.0	DAMP,73F	6880.	8050.	742.	7040.	-0
BBB 0	BRAND B1		3/4	3.10		.396	3.13	(28-DAY)	(56-DAY)	(28-DAY)	(28-DAY)	(28-DAY)
2/25/82	I	NONE	LIMESTONE	BRAND B1	NONE	.396	151.	• 7930. •	8640. •	861. •	7790. •	-0•
8.5/1.5	806.	0	BRAND B1	1090.	0	319.	75.0	8010.	8350.	842.	8120.	-0
	15.	0	1684.	25.		4.5	15. MIN	8010.	8480.	892.	7560.	-0
		0	39. 66.	1.54	-0	2.0	DAMP,73F	7780.	8680.	850.	7680.	-0
BBB 0	BRAND B1		3/4	3.10		.392	3.25	(28-DAY)	(56-DAY)	(28-DAY)	(28-DAY)	(28-DAY)
2/25/82	I	NONE	LIMESTONE	BRAND B1	NONE	.392	151.	• 7880. •	8820. •	794. •	7490. •	-0•
8.5/2.0	807.	0	BRAND B1	912.	0	316.	75.0	7750.	8730.	783.	7120.	-0
	15.	0	1871.	21.		4.4	15. MIN	8130.	8510.	808.	7880.	-0
		0	43. 73.	2.05	-0	2.0	DAMP,73F	7760.	9160.	792.	7480.	-0
CBB 0	BRAND C1		3/4	3.10		.369	4.25	(28-DAY)	(56-DAY)	(28-DAY)	(28-DAY)	(28-DAY)
3/ 1/82	II	NONE	LIMESTONE	BRAND B1	NONE	.369	150.	• 8110. •	7540. •	853. •	8790. •	-0•
8.5/1.0	819.	0	BRAND B1	1384.	0	303.	75.0	7800.	7530.	842.	8830.	-0
	15.	0	1419.	32.		4.2	15. MIN	8310.	7800.	858.	8750.	-0
		0	33. 55.	1.03	-0	2.0	DAMP,73F	8220.	7290.	858.	8790.	-0
CBB 0	BRAND C1		3/4	3.10		.362	3.13	(28-DAY)	(56-DAY)	(28-DAY)	(28-DAY)	(28-DAY)
3/ 1/82	II	NONE	LIMESTONE	BRAND B1	NONE	.362	152.	• 8120. •	7740. •	855. •	7500. •	-0•
8.5/1.5	820.	0	BRAND B1	1111.	0	297.	74.0	7940.	7320.	858.	7800.	-0
	15.	0	1709.	26.		4.1	15. MIN	8210.	7960.	875.	6490.	-0
		0	39. 67.	1.54	-0	2.0	DAMP,73F	8210.	7940.	833.	8200.	-0

	CEMENT	FLYASH	COARSE AGG	FINE AGG	ADMIXTURE	WATER	MISC	••••••••••••••• TEST RESULTS •••••••••••••••				
MIX I.D. MIX DATE CF/CAFA	BRAND TYPE LBS/CUYD PCT VOLUME	BRAND CLASS LBS/CUYD PCT VOLUME PCT REPLACED	SIZE MATERIAL SOURCE LBS/CUYD P:VOL,DRUW	FINENESS SOURCE LBS/CUYD PCT VOLUME CA/FA(LB/LB)	TYPE BRAND DOSE(OZ/100) (2ND TYPE) (2ND DOSE)	W/C W/B LB/CUYD GAL/SACK PCT AIR	SLUMP UNIT WT MIX TMP MX TIME CURING	6 X 12 CYLINDER (STEEL) (PSI)	6 X 12 CYLINDER (STEEL) (PSI)	6X6X18 BEAM (STEEL) (PSI)	4 X 8 CYLINDER (CARDRD) (PSI)	4 X 8 CYLINDER (STEEL) (PSI)
CBB 0 3/ 2/82 6.5/2.0	BRAND C1 II 819. 15.	NONE 0 0 0	3/4 LIMESTONE BRAND B1 1895. 43. 74.	3.10 BRAND B1 924. 21. 2.05	NONE 0 -0	.365 .365 299. 4.1 2.0	3.00 152. 77.0 15. MIN DAMP,73F	(28-DAY) • 8470. 8450. 8700. 8260.	(56-DAY) • 8400. 8810. 8100. 8290.	(28-DAY) • 989. 1050. 883. 1033.	(28-DAY) • 8610. 8970. 8360. 8590.	(28-DAY) • -0. -0 -0 -0
DBB 0 3/ 2/82 8.5/1.0	BRAND D1 I 808. 15.	NONE 0 0 0	3/4 LIMESTONE BRAND B1 1400. 32. 55.	3.10 BRAND B1 1360. 32. 1.03	NONE 0 -0	.399 .399 323. 4.5 2.0	3.50 149. 77.0 15. MIN DAMP,73F	(28-DAY) • 8540. 8950. 8080. 8580.	(56-DAY) • 8340. 8100. 8010. 8910.	(28-DAY) • 869. 858. 875. 875.	(28-DAY) • 8460. 8440. 8040. 8910.	(28-DAY) • -0. -0 -0 -0
DBB 0 3/ 3/82 8.5/1.6	BRAND D1 I 799. 15.	NONE 0 0 0	3/4 LIMESTONE BRAND B1 1661. 38. 65.	3.10 BRAND B1 1066. 25. 1.56	NONE 0 -0	.425 .425 340. 4.8 2.0	3.63 148. 75.2 15. MIN DAMP,73F	(28-DAY) • 7840. 8100. 7940. 7480.	(56-DAY) • 8370. 8380. 8360. 8360.	(28-DAY) • 886. 917. 892. 850.	(28-DAY) • 7800. 5170. 7680. 7920.	(28-DAY) • -0. -0 -0 -0
DBB 0 3/ 3/82 8.5/2.1	BRAND D1 I 806. 15.	NONE 0 0 0	3/4 LIMESTONE BRAND B1 1861. 43. 73.	3.10 BRAND B1 899. 21. 2.07	NONE 0 -0	.404 .404 325. 4.5 2.0	3.63 150. 74.3 15. MIN DAMP,73F	(28-DAY) • 8370. 7820. 8420. 8860.	(56-DAY) • 8170. 8080. 7850. 8580.	(28-DAY) • 939. 1008. 942. 867.	(28-DAY) • 6150. 5570. 5010. 7880.	(28-DAY) • -0. -0 -0 -0
EBB 0 3/ 4/82 8.5/1.0	BRAND E1 III 794. 15.	NONE 0 0 0	3/4 LIMESTONE BRAND B1 1373. 31. 54.	3.10 BRAND B1 1330. 31. 1.03	NONE 0 -0	.440 .440 350. 5.0 2.0	3.50 147. 75.2 15. MIN DAMP,73F	(28-DAY) • 8180. 8220. 8210. 8120.	(56-DAY) • 8100. 8580. 8130. 7590.	(28-DAY) • 950. 1050. 867. 933.	(28-DAY) • 8330. 8440. 8670. 7880.	(28-DAY) • -0. -0 -0 -0
EBB 0 3/ 4/82 8.5/1.5	BRAND E1 III 794. 15.	NONE 0 0 0	3/4 LIMESTONE BRAND B1 1653. 38. 64.	3.10 BRAND B1 1068. 25. 1.55	NONE 0 -0	.433 .433 344. 4.9 2.0	3.50 147. 74.3 15. MIN DAMP,73F	(28-DAY) • 8390. 8580. 8350. 8240.	(56-DAY) • 8370. 8630. 7820. 8650.	(28-DAY) • 922. 883. 958. 925.	(28-DAY) • 7730. 7800. 7760. 7640.	(28-DAY) • -0. -0 -0 -0
EBB 0 3/ 5/82 8.5/2.0	BRAND E1 III 791. 15.	NONE 0 0 0	3/4 LIMESTONE BRAND B1 1822. 42. 71.	3.10 BRAND B1 891. 21. 2.04	NONE 0 -0	.441 .441 348. 5.0 2.0	3.38 147. 74.3 15. MIN DAMP,73F	(28-DAY) • 8830. 8740. 9000. 8750.	(56-DAY) • 8000. 7220. 8970. 7800.	(28-DAY) • 941. 983. 933. 909.	(28-DAY) • 8170. 8280. 8360. 7880.	(28-DAY) • -0. -0 -0 -0
CBB 0 3/11/82 7.0/1.0	BRAND C1 II 652. 12.	NONE 0 0 0	3/4 LIMESTONE BRAND B1 1501. 34. 59.	3.10 BRAND B1 1465. 34. 1.02	NONE 0 -0	.448 .448 292. 5.0 2.0	3.00 150. 76.1 15. MIN DAMP,73F	(28-DAY) • 6940. 6900. 6850. 7070.	(56-DAY) • 7250. 7320. 7040. 7360.	(28-DAY) • 872. 867. 917. 833.	(28-DAY) • 7680. 7800. 7600. 7640.	(28-DAY) • -0. -0 -0 -0

	CEMENT	FLYASH	COARSE AGG	FINE AGG	ADMIXTURE	WATER	MISC	TEST RESULTS				
MIX I.D. MIX DATE CF/CAFA	BRAND TYPE LBS/CUYD PCT VOLUME	BRAND CLASS LBS/CUYD PCT VOLUME PCT REPLACED	SIZE MATERIAL SOURCE LBS/CUYD P:VOL,DRUW	FINENESS SOURCE LBS/CUYD PCT VOLUME CA/FA(LB/LB)	TYPE BRAND DOSE(OZ/100) (2ND TYPE) (2ND DOSE)	W/C W/B LB/CUYD GAL/SACK PCT AIR	SLUMP UNIT WT MIX TMP MX TIME CURING	6 X 12 CYLINDER (STEEL) (PSI)	6 X 12 CYLINDER (STEEL) (PSI)	6X6X18 BEAM (STEEL) (PSI)	4 X 8 CYLINDER (CARDBD) (PSI)	4 X 8 CYLINDER (STEEL) (PSI)
CBB 0	BRAND C1		3/4	3.10		.439	4.00	(28-DAY)	(56-DAY)	(28-DAY)	(28-DAY)	(28-DAY)
3/11/82	II	NONE	LIMESTONE	BRAND B1	NONE	.439	151.	• 6480. •	7180. •	794. •	6800. •	-0 •
7.0/1.5	653.	0	BRAND B1	1176.	0	287.	75.2	6770.	7110.	842.	6680.	-0
	12.	0	1807.	27.		5.0	15. MIN	6720.	7270.	888.	6760.	-0
		0	41. 70.	1.54	-0	2.0	DAMP,73F	5940.	7160.	733.	6960.	-0
CBB 0	BRAND C1		3/4	3.10		.458	3.50	(28-DAY)	(56-DAY)	(29-DAY)	(28-DAY)	(28-DAY)
3/12/82	II	NONE	LIMESTONE	BRAND B1	NONE	.458	152.	• 6420. •	6900. •	864. •	6900. •	-0 •
7.0/2.0	649.	0	BRAND B1	973.	0	298.	77.0	6220.	6830.	892.	6760.	-0
	12.	0	1987.	23.		5.2	15. MIN	6600.	6840.	850.	6760.	-0
		0	46. 77.	2.04	-0	2.0	DAMP,73F	6440.	7020.	850.	7080.	-0
CBB 0	BRAND C1		3/4	3.10		.332	4.50	(28-DAY)	(56-DAY)	(28-DAY)	(28-DAY)	(28-DAY)
3/12/82	II	NONE	LIMESTONE	BRAND B1	NONE	.332	151.	• 8150. •	8120. •	1061. •	8670. •	-0 •
10.0/1.0	988.	0	BRAND B1	1286.	0	328.	77.9	7600.	8680.	1075.	8360.	-0
	19.	0	1314.	30.		3.7	15. MIN	8450.	7500.	1050.	9030.	-0
		0	30. 51.	1.02	-0	2.0	DAMP,73F	8400.	8190.	1058.	8630.	-0
CBB 0	BRAND C1		3/4	3.10		.320	4.00	(28-DAY)	(56-DAY)	(28-DAY)	(28-DAY)	(28-DAY)
3/15/82	II	NONE	LIMESTONE	BRAND B1	NONE	.320	152.	• 8660. •	8060. •	1008. •	8330. •	-0 •
10.0/1.5	993.	0	BRAND B1	1030.	0	318.	77.0	8740.	8820.	1042.	8550.	-0
	19.	0	1595.	24.		3.6	15. MIN	8880.	7230.	1033.	8630.	-0
		0	37. 62.	1.55	-0	2.0	DAMP,73F	8360.	8130.	950.	7800.	-0
CBB 0	BRAND C1		3/4	3.10		.313	4.00	(28-DAY)	(56-DAY)	(28-DAY)	(28-DAY)	(28-DAY)
3/15/82	II	NONE	LIMESTONE	BRAND B1	NONE	.313	153.	• 8580. •	8870. •	955. •	8000. •	-0 •
10.0/2.1	996.	0	BRAND B1	863.	0	311.	77.0	8590.	8470.	950.	7960.	-0
	19.	0	1778.	20.		3.5	15. MIN	8560.	7670.	958.	8550.	-0
		0	41. 69.	2.06	-0	2.0	DAMP,73F	8590.	0	958.	7480.	-0
CCB 0	BRAND C1		1	3.10		.354	3.25	(28-DAY)	(56-DAY)	(28-DAY)	(28-DAY)	(28-DAY)
3/15/82	II	NONE	LIMESTONE	BRAND B1	NONE	.354	152.	• 8290. •	8070. •	906. •	7950. •	-0 •
8.5/1.0	826.	0	BRAND C1	1380.	0	292.	77.0	8220.	8060.	950.	7880.	-0
	16.	0	1433.	32.		4.0	15. MIN	8290.	8290.	975.	8280.	-0
		0	33. 54.	1.04	-0	2.0	DAMP,73F	8360.	7850.	792.	7680.	-0
CCB 0	BRAND C1		1	3.10		.351	3.75	(28-DAY)	(56-DAY)	(29-DAY)	(28-DAY)	(28-DAY)
3/15/82	II	NONE	LIMESTONE	BRAND B1	NONE	.351	153.	• 7730. •	8160. •	914. •	7400. •	-0 •
8.5/1.6	823.	0	BRAND C1	1104.	0	289.	75.2	7850.	8220.	950.	7600.	-0
	16.	0	1721.	26.		4.0	15. MIN	7520.	7960.	900.	7440.	-0
		0	40. 64.	1.56	-0	2.0	DAMP,73F	7830.	8310.	892.	7160.	-0
CCB 0	BRAND C1		1	3.10		.340	3.25	(28-DAY)	(56-DAY)	(28-DAY)	(28-DAY)	(28-DAY)
3/16/82	II	NONE	LIMESTONE	BRAND B1	NONE	.340	153.	• 8010. •	8050. •	972. •	8880. •	-0 •
8.5/2.1	827.	0	BRAND C1	926.	0	281.	76.1	8130.	7900.	967.	8550.	-0
	16.	0	1917.	21.		3.8	15. MIN	7850.	8190.	1000.	9390.	-0
		0	44. 72.	2.07	-0	2.0	DAMP,73F	8050.	8060.	950.	8700.	-0

MIX I.D. MIX DATE CF/CAFA	CEMENT BRAND TYPE LBS/CUYD PCT VOLUME	FLYASH BRAND CLASS LBS/CUYD PCT VOLUME PCT REPLACED	COARSE AGG SIZE MATERIAL SOURCE LBS/CUYD P:VOL,DRUW	FINE AGG FINENESS SOURCE LBS/CUYD PCT VOLUME CA/FA(LB/LB)	ADMIXTURE TYPE BRAND DOSE(OZ/100) (2ND TYPE) (2ND DOSE)	WATER W/C W/B LB/CUYD GAL/SACK PCT AIR	MISC SLUMP UNIT WT MIX TMP MX TIME CURING	.... TEST RESULTS .... 6 X 12 CYLINDER (STEEL) (PSI)	6 X 12 CYLINDER (STEEL) (PSI)	6X6X18 BEAM (STEEL) (PSI)	4 X 8 CYLINDER (CARDBD) (PSI)	4 X 8 CYLINDER (STEEL) (PSI)
CCB 0	BRAND C1		1	3.10		.425	3.00	(28-DAY)	(56-DAY)	(28-DAY)	(28-DAY)	(28-DAY)
3/16/82	II	NONE	LIMESTONE	BRAND B1	NONE	.425	150.	• 7410. •	7480. •	847. •	7530. •	-0.
7.0/1.0	658.	0	BRAND C1	1466.	0	279.	77.0	7060.	7220.	892.	7600.	-0
	12.	0	1516.	34.		4.8	15. MIN	7440.	5110.	925.	7600.	-0
		0	35. 57.	1.03	-0	2.0	DAMP,73F	7730.	7730.	725.	7400.	-0
CCB 0	BRAND C1		1	3.10		.397	3.25	(28-DAY)	(56-DAY)	(29-DAY)	(28-DAY)	(28-DAY)
3/16/82	II	NONE	LIMESTONE	BRAND B1	NONE	.397	152.	• 7230. •	7410. •	869. •	7670. •	-0.
7.0/1.5	664.	0	BRAND C1	1183.	0	264.	77.0	7320.	7320.	833.	7720.	-0
	13.	0	1837.	27.		4.5	15. MIN	7370.	7500.	925.	7680.	-0
		0	42. 69.	1.55	-0	2.0	DAMP,73F	7000.	0	850.	7600.	-0
CCB 0	BRAND C1		1	3.10		.419	4.00	(28-DAY)	(56-DAY)	(28-DAY)	(28-DAY)	(28-DAY)
3/16/82	II	NONE	LIMESTONE	BRAND B1	NONE	.419	152.	• 6960. •	7100. •	845. •	7190. •	-0.
7.0/2.1	657.	0	BRAND C1	976.	0	275.	78.8	6970.	6930.	817.	7080.	-0
	12.	0	2020.	23.		4.7	15. MIN	6970.	7020.	842.	7320.	-0
		0	47. 76.	2.07	-0	2.0	DAMP,73F	6930.	7340.	875.	7160.	-0
CCB 0	BRAND C2		1	3.10		.315	3.50	(28-DAY)	(56-DAY)	(28-DAY)	(28-DAY)	(28-DAY)
3/24/82	II	NONE	LIMESTONE	BRAND B1	NONE	.315	151.	• 8340. •	9140. •	997. •	8310. •	-0.
10.0/1.0	992.	0	BRAND C1	1294.	0	312.	77.0	8310.	9580.	958.	8280.	-0
	19.	0	1332.	30.		3.5	15. MIN	7960.	9180.	1050.	8280.	-0
		0	31. 50.	1.03	-0	2.0	DAMP,73F	8740.	8670.	983.	8360.	-0
CCB 0	BRAND C2		1	3.10		.334	4.00	(28-DAY)	(56-DAY)	(28-DAY)	(28-DAY)	(28-DAY)
3/26/82	II	NONE	LIMESTONE	BRAND B1	NONE	.334	151.	• 7450. •	7990. •	928. •	7550. •	-0.
16.0/1.5	979.	0	BRAND C1	1019.	0	327.	74.3	7750.	8260.	1042.	7520.	-0
	18.	0	1582.	24.		3.8	15. MIN	7160.	7780.	925.	8040.	-0
		0	37. 59.	1.55	-0	2.0	DAMP,73F	7440.	7640.	817.	7080.	-0
CCB 0	BRAND C2		1	3.10		.327	4.00	(28-DAY)	(56-DAY)	(29-DAY)	(28-DAY)	(28-DAY)
4/ 5/82	II	NONE	LIMESTONE	BRAND B1	NONE	.327	150.	• 6830. •	7810. •	930. •	7490. •	-0.
10.0/2.1	980.	0	BRAND C1	847.	0	320.	77.0	6830.	7850.	917.	7600.	-0
	18.	0	1770.	20.		3.7	15. MIN	7160.	7920.	942.	7120.	-0
		0	41. 66.	2.09	-0	2.0	DAMP,73F	6510.	7670.	0	7760.	-0
ACB 0	BRAND A1		1	3.10		.375	3.75	(28-DAY)	(56-DAY)	(28-DAY)	(28-DAY)	(28-DAY)
3/17/82	I	NONE	LIMESTONE	BRAND B1	NONE	.375	150.	• 7740. •	8210. •	975. •	8120. •	-0.
8.5/1.0	815.	0	BRAND C1	1365.	0	306.	76.1	7500.	8450.	917.	8120.	-0
	15.	0	1421.	32.		4.2	15. MIN	8060.	7940.	975.	8000.	-0
		0	33. 53.	1.04	-0	2.0	DAMP,73F	7660.	8240.	1033.	8240.	-0
ACB 0	BRAND A1		1	3.10		.357	4.25	(28-DAY)	(56-DAY)	(28-DAY)	(28-DAY)	(28-DAY)
3/17/82	I	NONE	LIMESTONE	BRAND B1	NONE	.357	151.	• 7590. •	7850. •	972. •	7360. •	-0.
8.5/2.1	819.	0	BRAND C1	914.	0	293.	76.1	7750.	8080.	975.	7480.	-0
	15.	0	1906.	21.		4.0	15. MIN	7550.	8010.	1017.	7320.	-0
		0	44. 71.	2.09	-0	2.0	DAMP,73F	7480.	7460.	925.	7280.	-0

	CEMENT	FLYASH	COARSE AGG	FINE AGG	ADMIXTURE	WATER	MISC	************** TEST RESULTS				**************
MIX I.D. MIX DATE CF/CAFA	BRAND TYPE LBS/CUYD PCT VOLUME	BRAND CLASS LBS/CUYD PCT VOLUME PCT REPLACED	SIZE MATERIAL SOURCE LBS/CUYD P:VOL,DRUW	FINENESS SOURCE LBS/CUYD PCT VOLUME CA/FA(LB/LB)	TYPE BRAND DOSE(OZ/100) (2ND TYPE) (2ND DOSE)	W/C W/B LB/CUYD GAL/SACK PCT AIR	SLUMP UNIT WT MIX TMP MX TIME CURING	6 X 12 CYLINDER (STEEL) (PSI)	6 X 12 CYLINDER (STEEL) (PSI)	6X6X18 BEAM (STEEL) (PSI)	4 X 8 CYLINDER (CARDBD) (PSI)	4 X 8 CYLINDER (STEEL) (PSI)
COB 0 4/ 5/82 8.5/1.0	BRAND C2 II 808. 15.	NONE 0 0 0	1/2 LIMESTONE BRAND D1 1371. 33. 60.	3.10 BRAND B1 1327. 31. 1.03	NONE 0 -0	.395 .395 319. 4.4 2.0	3.50 147. 74.3 15. MIN DAMP,73F	(28-DAY) * 7360. 7500. 7390. 7180.	(56-DAY) * 8030. 8150. 7640. 8310.	(28-DAY) * 825. 850. 825. 800.	(28-DAY) * 7280. 7640. 7040. 7160.	(28-DAY) * -0* -0 -0 -0
COB 0 4/ 6/82 8.5/1.6	BRAND C2 II 797. 15.	NONE 0 0 0	1/2 LIMESTONE BRAND D1 1612. 39. 70.	3.10 BRAND B1 998. 23. 1.61	NONE 0 -0	.443 .443 353. 5.0 2.0	3.25 146. 72.5 15. MIN DAMP,73F	(28-DAY) * 7120. 6740. 7390. 7230.	(56-DAY) * 7730. 7820. 7670. 7690.	(28-DAY) * 786. 800. 783. 775.	(28-DAY) * 7410. 7320. 7520. 7400.	(28-DAY) * -0* -0 -0 -0
COB 0 4/ 6/82 8.5/2.2	BRAND C2 II 795. 15.	NONE 0 0 0	1/2 LIMESTONE BRAND D1 1785. 43. 78.	3.10 BRAND B1 828. 19. 2.16	NONE 0 -0	.440 .440 350. 5.0 2.0	3.25 147. 71.6 15. MIN DAMP,73F	(28-DAY) * 6870. 6420. 7270. 6910.	(56-DAY) * 7560. 7390. 7660. 7640.	(28-DAY) * 764. 733. 825. 733.	(28-DAY) * 6550. 6760. 6410. 6490.	(28-DAY) * -0* -0 -0 -0
COB 0 4/ 7/82 7.0/1.1	BRAND C2 II 637. 12.	NONE 0 0 0	1/2 LIMESTONE BRAND D1 1428. 34. 62.	3.10 BRAND B1 1328. 31. 1.08	NONE 0 -0	.549 .549 350. 6.2 2.0	3.25 146. 75.2 15. MIN DAMP,73F	(28-DAY) * 5770. 5660. 5730. 5910.	(56-DAY) * 6720. 7040. 6670. 6440.	(28-DAY) * 639. 650. 650. 617.	(28-DAY) * 5290. 6050. 4850. 4970.	(28-DAY) * -0* -0 -0 -0
COB 0 4/ 7/82 7.0/1.6	BRAND C2 II 644. 12.	NONE 0 0 0	1/2 LIMESTONE BRAND D1 1726. 42. 75.	3.10 BRAND B1 1070. 25. 1.61	NONE 0 -0	.508 .508 327. 5.7 2.0	3.75 146. 74.3 15. MIN DAMP,73F	(28-DAY) * 5870. 5710. 5890. 6010.	(56-DAY) * 6790. 7070. 6540. 6770.	(28-DAY) * 664. 675. 633. 683.	(28-DAY) * 5110. 5170. 5050. 3180.	(28-DAY) * -0* -0 -0 -0
COB 0 4/15/82 7.0/2.0	BRAND C2 II 634. 12.	NONE 0 0 0	1/2 LIMESTONE BRAND D1 1887. 46. 82.	3.10 BRAND B1 919. 21. 2.05	NONE 0 -0	.511 .511 324. 5.8 2.0	3.50 147. 77.0 15. MIN DAMP,73F	(28-DAY) * 5610. 5710. 5640. 5480.	(56-DAY) * 6710. 6860. 6680. 6600.	(28-DAY) * 733. 733. 733. 733.	(28-DAY) * 5740. 5490. 5970. 5770.	(28-DAY) * -0* -0 -0 -0
COB 0 4/20/82 10.0/1.1 (U)	BRAND C2 II 988. 19.	NONE 0 0 0	1/2 LIMESTONE BRAND D1 1290. 31. 56.	3.10 BRAND B1 1205. 28. 1.07	NONE 0 -0	.346 .346 342. 3.9 2.0	3.00 147. 77.0 15. MIN DAMP,73F	(28-DAY) * 7840. 7820. 7800. 7900.	(56-DAY) * 8750. 8490. 8560. 9200.	(28-DAY) * 889. 875. 958. 833.	(28-DAY) * 8330. 8710. 8280. 8000.	(28-DAY) * -0* -0 -0 -0
COB 0 4/20/82 10.0/1.6	BRAND C2 II 982. 18.	NONE 0 0 0	1/2 LIMESTONE BRAND D1 1531. 37. 67.	3.10 BRAND B1 954. 22. 1.60	NONE 0 -0	.351 .351 345. 4.0 2.0	3.00 147. 75.2 15. MIN DAMP,73F	(28-DAY) * 7910. 7670. 8510. 7550.	(56-DAY) * 8630. 8930. 8260. 8700.	(28-DAY) * 803. 842. 767. 800.	(28-DAY) * 8410. 8670. 8360. 8200.	(28-DAY) * -0* -0 -0 -0

MIX I.D. MIX DATE CF/CAFA	CEMENT BRAND TYPE LBS/CUYD PCT VOLUME	FLYASH BRAND CLASS LBS/CUYD PCT VOLUME PCT REPLACED	COARSE AGG SIZE MATERIAL SOURCE LBS/CUYD P:VOL,DRUM	FINE AGG FINENESS SOURCE LBS/CUYD PCT VOLUME CA/FA(LB/LB)	ADMIXTURE TYPE BRAND DOSE(OZ/100) (2ND TYPE) (2ND DOSE)	WATER W/C W/B LB/CUYD GAL/SACK PCT AIR	MISC SLUMP UNIT WT MIX TMP MX TIME CURING	TEST RESULTS 6 X 12 CYLINDER (STEEL) (PSI)	TEST RESULTS 6 X 12 CYLINDER (STEEL) (PSI)	TEST RESULTS 6X6X18 BEAM (STEEL) (PSI)	TEST RESULTS 4 X 8 CYLINDER (CARDBD) (PSI)	TEST RESULTS 4 X 8 CYLINDER (STEEL) (PSI)
CDB 0	BRAND C2		1/2	3.10		.379	3.75	(28-DAY)	(56-DAY)	(28-DAY)	(28-DAY)	(28-DAY)
4/20/82	II	NONE	LIMESTONE	BRAND B1	NONE	.379	145.	6920.	7640.	772.	7240.	-0.
16.0/2.1	963.	0	BRAND D1	777.	0	365.	75.2	6680.	8030.	742.	7000.	-0
	18.	0	1665.	18.		4.3	15. MIN	7020.	7070.	733.	7400.	-0
		0	40. 72.	2.14	-0	2.0	DAMP,73F	7070.	7930.	842.	7320.	-0
ADB 0	BRAND A1		1/2	3.10		.385	3.00	(28-DAY)	(56-DAY)	(28-DAY)	(28-DAY)	(28-DAY)
3/17/82	I	NONE	LIMESTONE	BRAND B1	NONE	.385	148.	7850.	8020.	911.	8120.	-0.
8.5/1.0	814.	0	BRAND D1	1331.	0	313.	77.0	7780.	7480.	875.	8000.	-0
	15.	0	1376.	31.		4.3	15. MIN	7750.	7990.	908.	8360.	-0
		0	33. 60.	1.03	-0	2.0	DAMP,73F	8030.	8590.	950.	8000.	-0
ADB 0	BRAND A1		1/2	3.10		.400	3.00	(28-DAY)	(56-DAY)	(28-DAY)	(28-DAY)	(28-DAY)
3/17/82	I	NONE	LIMESTONE	BRAND B1	NONE	.400	148.	7360.	7390.	881.	7310.	-0.
8.5/2.1	804.	0	BRAND D1	873.	0	322.	77.0	7460.	7460.	817.	7240.	-0
	15.	0	1803.	20.		4.5	15. MIN	7230.	7290.	908.	7200.	-0
		0	43. 78.	2.07	-0	2.0	DAMP,73F	7390.	7430.	917.	7480.	-0
CCB 0	BRAND C2		1/2	3.10		.386	7.50	(28-DAY)	(56-DAY)	(28-DAY)	(28-DAY)	(28-DAY)
4/15/82	II	NONE	LIMESTONE	BRAND B1	NONE	.386	147.	6890.	7440.	847.	6970.	-0.
10.0/1.0	954.	0	BRAND D1	1213.	0	369.	77.0	6860.	7690.	858.	6920.	-0
(U)	18.	0	1244.	28.		4.4	15. MIN	6740.	7180.	875.	7120.	-0
		0	30. 54.	1.03	-0	2.0	DAMP,73F	7070.	5500.	808.	6880.	-0
CEB 0	BRAND C2		1/2	3.10		.492	3.75	(28-DAY)	(56-DAY)	(28-DAY)	(28-DAY)	(28-DAY)
5/17/82	II	NONE	LIMESTONE	BRAND B1	NONE	.492	148.	6070.	6470.	803.	6500.	-0.
7.0/1.0	641.	0	BRAND E1	1449.	0	315.	78.8	5920.	6510.	800.	6530.	-0
	12.	0	1501.	34.		5.5	15. MIN	6050.	6120.	800.	6490.	-0
		0	34. 57.	1.04	-0	2.0	DAMP,73F	6240.	6770.	808.	6490.	-0
CEB 0	BRAND C2		1/2	3.10		.478	3.00	(28-DAY)	(56-DAY)	(28-DAY)	(28-DAY)	(28-DAY)
5/17/82	II	NONE	LIMESTONE	BRAND B1	NONE	.478	151.	5760.	6760.	797.	6560.	-0.
7.0/1.5	643.	0	BRAND E1	1166.	0	307.	78.8	4880.	6930.	825.	6600.	-0
	12.	0	1813.	27.		5.4	15. MIN	6470.	6880.	767.	6600.	-0
		0	41. 69.	1.55	-0	2.0	DAMP,73F	5920.	6470.	800.	6490.	-0
CEB 0	BRAND C2		1/2	3.10		.528	3.25	(28-DAY)	(56-DAY)	(28-DAY)	(28-DAY)	(28-DAY)
5/19/82	II	NONE	LIMESTONE	BRAND B1	NONE	.528	149.	5830.	6910.	847.	5930.	-0.
7.0/2.0	626.	0	BRAND E1	984.	0	331.	77.9	5940.	6610.	850.	6130.	-0
	12.	0	1954.	23.		6.0	15. MIN	5850.	7020.	850.	5670.	-0
		0	44. 75.	1.99	-0	2.0	DAMP,73F	5710.	7110.	842.	6000.	-0
CEB 0	BRAND C2		1/2	3.10		.392	3.50	(28-DAY)	(56-DAY)	(28-DAY)	(28-DAY)	(28-DAY)
5/19/82	II	NONE	LIMESTONE	BRAND B1	NONE	.392	149.	7360.	8520.	900.	7680.	-0.
8.5/1.0	802.	0	BRAND E1	1415.	0	314.	77.9	7360.	8720.	867.	7990.	-0
	15.	0	1404.	33.		4.4	15. MIN	7640.	8400.	933.	7890.	-0
		0	31. 54.	.99	-0	2.0	DAMP,73F	7070.	8440.	900.	7160.	-0

MIX I.D. MIX DATE CF/CAFA	CEMENT BRAND TYPE LBS/CUYD PCT VOLUME	FLYASH BRAND CLASS LBS/CUYD PCT VOLUME PCT REPLACED	COARSE AGG SIZE MATERIAL SOURCE LBS/CUYD P:VOL,DRUM	FINE AGG FINENESS SOURCE LBS/CUYD PCT VOLUME CA/FA(LB/LB)	ADMIXTURE TYPE BRAND DOSE(OZ/100) (2ND TYPE) (2ND DOSE)	WATER W/C W/B LB/CUYD GAL/SACK PCT AIR	MISC SLUMP UNIT WT MIX TMP MX TIME CURING	TEST RESULTS 6 X 12 CYLINDER (STEEL) (PSI)	TEST RESULTS 6 X 12 CYLINDER (STEEL) (PSI)	TEST RESULTS 6X6X18 BEAM (STEEL) (PSI)	TEST RESULTS 4 X 8 CYLINDER (CARDBD) (PSI)	TEST RESULTS 4 X 8 CYLINDER (STEEL) (PSI)
CEB 0 5/20/82 8.5/1.5	BRAND C2 II 805. 15.	NONE 0 0 0	1/2 LIMESTONE BRAND E1 1691. 38. 65.	3.10 BRAND B1 1121. 26. 1.51	NONE 0 -0	.397 .397 320. 4.5 2.0	3.50 151. 77.9 15. MIN DAMP,73F	(28-DAY) 7560. 7200. 7750. 7730.	(56-DAY) 8510. 8790. 8520. 8510.	(28-DAY) 906. 875. 925. 917.	(28-DAY) 7860. 7620. 8120. 7830.	(28-DAY) -0. -0 -0 -0
CLB 0 5/20/82 8.5/2.0	BRAND C2 II 803. 15.	NONE 0 0 0	1/2 LIMESTONE BRAND E1 1881. 42. 72.	3.10 BRAND B1 934. 22. 2.01	NONE 0 -0	.400 .400 322. 4.5 2.0	4.00 151. 78.8 15. MIN DAMP,73F	(28-DAY) 7770. 7570. 7920. 7820.	(56-DAY) 8670. 8990. 8740. 8280.	(28-DAY) 922. 933. 967. 867.	(28-DAY) 7880. 7830. 8090. 7730.	(28-DAY) -0. -0 -0 -0
CLB 0 5/25/82 10.0/1.0	BRAND C2 II 948. 18.	NONE 0 0 0	1/2 LIMESTONE BRAND E1 1343. 30. 51.	3.10 BRAND B1 1328. 31. 1.01	NONE 0 -0	.343 .343 325. 3.9 2.0	3.75 149. 78.8 15. MIN DAMP,73F	(28-DAY) 8420. 8540. 8290. 8440.	(56-DAY) 9220. 9200. 9550. 8910.	(28-DAY) 975. 975. 992. 958.	(28-DAY) 8340. 8210. 8940. 7880.	(28-DAY) -0. -0 -0 -0
CLB 0 5/27/82 10.0/1.6 (V)	BRAND C2 II 959. 18.	NONE 0 0 0	1/2 LIMESTONE BRAND E1 1639. 37. 63.	3.10 BRAND B1 1005. 23. 1.63	NONE 0 -0	.350 .350 336. 3.9 2.0	4.25 150. 79.7 15. MIN DAMP,73F	(28-DAY) 8130. 7900. 8360. 6300.	(56-DAY) 9010. 9230. 8010. 8790.	(28-DAY) 989. 942. 958. 1067.	(28-DAY) 0 0 0 0	(28-DAY) -0. -0 -0 -0
CLB 0 5/26/82 10.0/2.0 (Q)	BRAND C2 II 960. 18.	NONE 0 0 0	1/2 LIMESTONE BRAND E1 1823. 41. 70.	3.10 BRAND B1 896. 21. 2.03	NONE 0 -0	.322 .322 309. 3.6 2.0	3.00 152. 79.7 15. MIN DAMP,73F	(28-DAY) 8430. 8590. 8630. 8060.	(56-DAY) 9560. 9600. 9480. 9600.	(28-DAY) 1039. 1017. 1050. 1050.	(28-DAY) 8600. 8850. 8440. 8500.	(28-DAY) -0. -0 -0 -0
CFB 0 5/26/82 10.0/1.5 (V)	BRAND C2 II 939. 18.	NONE 0 0 0	1/2 LIMESTONE BRAND E1 1602. 36. 61.	3.10 BRAND B1 1048. 24. 1.53	NONE 0 -0	.361 .361 339. 4.1 2.0	5.25 150. 79.7 15. MIN DAMP,73F	(28-DAY) 8220. 8080. 8240. 8350.	(56-DAY) 9000. 9040. 8860. 9110.	(28-DAY) 1017. 1025. 992. 1033.	(28-DAY) 8360. 8360. 0 0	(28-DAY) -0. -0 -0 -0
CLB 0 5/25/82 10.0/1.5 (V)	BRAND C2 II 943. 18.	NONE 0 0 0	1/2 LIMESTONE BRAND E1 1609. 36. 61.	3.10 BRAND B1 1059. 25. 1.52	NONE 0 -0	.351 .351 331. 4.0 2.0	5.50 150. 78.8 15. MIN DAMP,73F	(28-DAY) 7810. 7570. 8050. 7800.	(56-DAY) 9080. 9000. 9090. 9140.	(28-DAY) 940. 1030. 958. 833.	(28-DAY) 7840. 7260. 8100. 8160.	(28-DAY) -0. -0 -0 -0
CLC 0 5/27/82 7.0/1.0	BRAND C3 II 667. 13.	NONE 0 0 0	1/2 LIMESTONE BRAND E1 1533. 34. 59.	2.72 BRAND C1 1512. 34. 1.01	NONE 0 -0	.425 .425 284. 4.8 2.0	3.00 152. 77.0 15. MIN DAMP,73F	(28-DAY) 7130. 7200. 6880. 7320.	(56-DAY) 7460. 7070. 7600. 7710.	(28-DAY) 894. 908. 858. 917.	(28-DAY) -0 -0 -0 -0	(28-DAY) -0. -0 -0 -0

MIX I.D. MIX DATE CF/CAFA	CEMENT BRAND TYPE LBS/CUYD PCT VOLUME	FLYASH BRAND CLASS LBS/CUYD PCT VOLUME PCT REPLACED	COARSE AGG SIZE MATERIAL SOURCE LBS/CUYD P:VOL,DRUW	FINE AGG FINENESS SOURCE LBS/CUYD PCT VOLUME CA/FA(LB/LB)	ADMIXTURE TYPE BRAND DOSE(OZ/100) (2ND TYPE) (2ND DOSE)	WATER W/C W/B LB/CUYD GAL/SACK PCT AIR	MISC SLUMP UNIT WT MIX TMP MX TIME CURING	TEST RESULTS 6 X 12 CYLINDER (STEEL) (PSI)	6 X 12 CYLINDER (STEEL) (PSI)	6X6X18 BEAM (STEEL) (PSI)	4 X 8 CYLINDER (CARDBD) (PSI)	4 X 8 CYLINDER (STEEL) (PSI)
CEC 0 5/28/82 7.0/1.5	BRAND C3 II 662. 12.	 NONE 0 0 0	1/2 LIMESTONE BRAND E1 1821. 41. 70.	2.72 BRAND C1 1200. 27. 1.52	 NONE 0 -0	.446 .446 296. 5.0 2.0	3.25 152. 78.8 15. MIN DAMP,73F	(28-DAY) • 7270. 7360. 7230. 7220.	(56-DAY) • 8190. 7990. 8380. 8210.	(28-DAY) • 928. 917. 958. 908.	(28-DAY) • -0 -0 -0 -0	(28-DAY) • -0• -0 -0 -0
CEC 0 5/28/82 7.0/2.0	BRAND C3 II 658. 12.	 NONE 0 0 0	1/2 LIMESTONE BRAND E1 2012. 45. 77.	2.72 BRAND C1 994. 23. 2.02	 NONE 0 -0	.461 .461 304. 5.2 2.0	3.00 152. 79.7 15. MIN DAMP,73F	(28-DAY) • 6910. 6630. 7140. 6970.	(56-DAY) • 7310. 6790. 7360. 7780.	(28-DAY) • 886. 842. 917. 900.	(28-DAY) • -0 -0 -0 -0	(28-DAY) • -0• -0 -0 -0
CEC 0 5/31/82 8.5/1.0	BRAND C3 II 789. 15.	 NONE 0 0 0	1/2 LIMESTONE BRAND E1 1458. 33. 56.	2.72 BRAND C1 1431. 32. 1.02	 NONE 0 -0	.386 .386 304. 4.3 2.0	3.50 152. 77.9 15. MIN DAMP,73F	(28-DAY) • 8260. 8740. 7640. 8400.	(56-DAY) • 8660. 8220. 8450. 9300.	(28-DAY) • 975. 983. 917. 1025.	(28-DAY) • 8730. 8420. 8990. 8790.	(28-DAY) • -0• -0 -0 -0
CEC 0 5/31/82 8.5/1.5	BRAND C3 II 795. 15.	 NONE 0 0 0	1/2 LIMESTONE BRAND E1 1766. 40. 67.	2.72 BRAND C1 1156. 26. 1.53	 NONE 0 -0	.366 .366 291. 4.1 2.0	3.25 154. 77.9 15. MIN DAMP,73F	(28-DAY) • 8010. 7850. 7980. 8210.	(56-DAY) • 9160. 9230. 9140. 9120.	(28-DAY) • 922. 942. 875. 950.	(28-DAY) • 8080. 8510. 8100. 7640.	(28-DAY) • -0• -0 -0 -0
CEC 0 6/ 1/82 8.5/2.0	BRAND C3 II 793. 15.	 NONE 0 0 0	1/2 LIMESTONE BRAND E1 1952. 44. 75.	2.72 BRAND C1 959. 22. 2.04	 NONE 0 -0	.374 .374 297. 4.2 2.0	3.00 153. 77.0 15. MIN DAMP,73F	(28-DAY) • 8260. 8350. 8290. 8130.	(58-DAY) • 8610. 8490. 9300. 8030.	(28-DAY) • 984. 992. 967. 992.	(28-DAY) • 8290. 7810. 8790. 8280.	(28-DAY) • -0• -0 -0 -0
CEC 0 6/ 1/82 10.0/1.0	BRAND C3 II 932. 18.	 NONE 0 0 0	1/2 LIMESTONE BRAND E1 1399. 31. 53.	2.72 BRAND C1 1376. 31. 1.02	 NONE 0 -0	.325 .325 303. 3.7 2.0	3.25 152. 77.9 15. MIN DAMP,73F	(28-DAY) • 8870. 8220. 8950. 9430.	(58-DAY) • 9440. 9600. 9690. 9020.	(28-DAY) • 1008. 975. 1067. 983.	(28-DAY) • 8810. 8880. 8370. 9170.	(28-DAY) • -0• -0 -0 -0
CEC 0 6/ 2/82 10.0/1.5	BRAND C3 II 928. 17.	 NONE 0 0 0	1/2 LIMESTONE BRAND E1 1676. 38. 64.	2.72 BRAND C1 1098. 25. 1.53	 NONE 0 -0	.328 .328 305. 3.7 2.0	3.00 152. 77.9 15. MIN DAMP,73F	(28-DAY) • 8810. 9110. 8060. 9270.	(57-DAY) • 9370. 9690. 9530. 8890.	(28-DAY) • 989. 1058. 1025. 883.	(28-DAY) • 9530. 9220. 9710. 9660.	(28-DAY) • -0• -0 -0 -0
CEC 0 6/ 2/82 10.0/2.0	BRAND C3 II 933. 18.	 NONE 0 0 0	1/2 LIMESTONE BRAND E1 1873. 42. 72.	2.72 BRAND C1 920. 21. 2.04	 NONE 0 -0	.318 .318 297. 3.6 2.0	3.25 154. 77.9 15. MIN DAMP,73F	(28-DAY) • 8580. 8420. 8610. 8720.	(57-DAY) • 9410. 9350. 9400. 9280.	(28-DAY) • 1036. 1050. 992. 1067.	(28-DAY) • 9310. 9570. 8910. 9450.	(28-DAY) • -0• -0 -0 -0

MIX I.D. MIX DATE CF/CAFA	CEMENT BRAND TYPE LBS/CUYD PCT VOLUME	FLYASH BRAND CLASS LBS/CUYD PCT VOLUME PCT REPLACED	COARSE AGG SIZE MATERIAL SOURCE LBS/CUYD P:VOL,DRUW	FINE AGG FINENESS SOURCE LBS/CUYD PCT VOLUME CA/FA(LB/LB)	ADMIXTURE TYPE BRAND DOSE(OZ/100) (2ND TYPE) (2ND DOSE)	WATER W/C W/B LB/CUYD GAL/SACK PCT AIR	MISC SLUMP UNIT WT MIX TMP MX TIME CURING	TEST RESULTS 6 X 12 CYLINDER (STEEL) (PSI)	6 X 12 CYLINDER (STEEL) (PSI)	6X6X18 BEAM (STEEL) (PSI)	4 X 8 CYLINDER (CAPDBD) (PSI)	4 X 8 CYLINDER (STEEL) (PSI)
EEC 0 6/ 3/82 10.0/1.0	BRAND E1 III 899. 17.	NONE 0 0 0	1/2 LIMESTONE BRAND E1 1350. 30. 52.	2.72 BRAND C1 1321. 30. 1.02	NONE 0 -0	.392 .392 352. 4.4 2.0	3.25 148. 80.6 15. MIN DAMP,73F	(28-DAY) • 8200. • 8820. 7660. 8130.	(58-DAY) 8670. • 8310. 8590. 9120.	(28-DAY) 1039. • 1017. 1058. 1042.	(28-DAY) 7790. • 8120. 7500. 7720.	(28-DAY) -0• -0 -0 -0
FEC 0 6/ 3/82 10.0/1.5	BRAND E1 III 934. 18.	NONE 0 0 0	1/2 LIMESTONE BRAND E1 1594. 36. 51.	2.72 BRAND C1 1040. 24. 1.53	NONE 0 -0	.381 .381 356. 4.3 2.0	3.25 148. 80.6 15. MIN DAMP,73F	(28-DAY) • 7740. • 8050. 7230. 7940.	(58-DAY) 8710. • 8790. 8580. 8750.	(28-DAY) 944. • 933. 958. 942.	(28-DAY) 7710. • 7750. 8050. 7340.	(28-DAY) -0• -0 -0 -0
EEC 0 7/16/82 8.5/1.5	BRAND E1 III 772. 15.	NONE 0 0 0	1/2 LIMESTONE BRAND E2 1713. 39. 67.	2.72 BRAND C1 1127. 26. 1.52	NONE 0 -0	.424 .424 327. 4.8 2.0	3.25 148. 80.6 15. MIN DAMP,73F	(28-DAY) • 8020. • 7830. 8400. 7820.	(56-DAY) 8550. • 8540. 8940. 8280.	(29-DAY) 897. • 858. 942. 892.	(28-DAY) 7720. • 7790. 7480. 7900.	(28-DAY) -0• -0 -0 -0
DEC 0 7/19/82 10.0/1.5	BRAND D1 I 911. 17.	NONE 0 0 0	1/2 LIMESTONE BRAND E2 1659. 37. 65.	2.72 BRAND C1 1085. 25. 1.53	NONE 0 -0	.351 .351 320. 4.0 2.0	4.25 149. 80.6 15. MIN DAMP,73F	(28-DAY) • 7980. • 8050. 7750. 8130.	(56-DAY) 8960. • 8750. 9070. 9070.	(28-DAY) 900. • 883. 917. 900.	(28-DAY) 8180. • 8020. 8280. 8250.	(28-DAY) -0• -0 -0 -0
DEC 0 7/19/82 8.5/1.5	BRAND D1 I 776. 15.	NONE 0 0 0	1/2 LIMESTONE BRAND E2 1737. 39. 68.	2.72 BRAND C1 1134. 26. 1.53	NONE 0 -0	.405 .405 314. 4.6 2.0	3.50 148. 80.6 15. MIN DAMP,73F	(28-DAY) • 8040. • 7820. 8260. 8030.	(56-DAY) 8290. • 8880. 8240. 8560.	(28-DAY) 889. • 933. 875. 858.	(28-DAY) 7930. • 7970. 7940. 7870.	(28-DAY) -0• -0 -0 -0
CED 0 6/ 7/82 7.0/1.0	BRAND C3 II 673. 13.	NONE 0 0 0	1/2 LIMESTONE BRAND E2 1516. 34. 59.	2.81 BRAND D1 1476. 33. 1.03	NONE 0 -0	.446 .446 300. 5.0 2.0	3.25 149. 77.0 15. MIN DAMP,73F	(28-DAY) • 6930. • 6830. 6880. 7090.	(56-DAY) 7420. • 7710. 7590. 6950.	(28-DAY) 828. • 833. 833. 817.	(28-DAY) 6620. • 6130. 6460. 7260.	(28-DAY) -0• -0 -0 -0
CED 0 6/ 7/82 7.0/1.5	BRAND C3 II 672. 13.	NONE 0 0 0	1/2 LIMESTONE BRAND E2 1819. 41. 71.	2.81 BRAND D1 1180. 27. 1.54	NONE 0 -0	.444 .444 298. 5.0 2.0	3.50 151. 77.9 15. MIN DAMP,73F	(28-DAY) • 6910. • 7160. 6510. 7070.	(56-DAY) 7560. • 7460. 8080. 7140.	(29-DAY) 841. • 858. 833. 833.	(28-DAY) 6650. • 7000. 6110. 6940.	(28-DAY) -0• -0 -0 -0
CED 0 6/ 8/82 7.0/2.1	BRAND C3 II 668. 13.	NONE 0 0 0	1/2 LIMESTONE BRAND E2 2005. 45. 78.	2.81 BRAND D1 972. 22. 2.06	NONE 0 -0	.462 .462 309. 5.2 2.0	3.25 151. 77.9 15. MIN DAMP,73F	(28-DAY) • 6440. • 6470. 6540. 6300.	(56-DAY) 7320. • 7360. 7360. 7250.	(29-DAY) 803. • 808. 767. 833.	(28-DAY) 6290. • 6750. 5510. 6620.	(28-DAY) -0• -0 -0 -0

MIX I.D. MIX DATE CF/CAFA	CEMENT BRAND TYPE LBS/CUYD PCT VOLUME	FLYASH BRAND CLASS LBS/CUYD PCT VOLUME PCT REPLACED	COARSE AGG SIZE MATERIAL SOURCE LBS/CUYD P:VOL,DRUM	FINE AGG FINENESS SOURCE LBS/CUYD PCT VOLUME CA/FA(LB/LB)	ADMIXTURE TYPE BRAND DOSE(OZ/100) (2ND TYPE) (2ND DOSE)	WATER W/C W/B LB/CUYD GAL/SACK PCT AIR	MISC SLUMP UNIT WT MIX TMP MX TIME CURING	TEST RESULTS 6 X 12 CYLINDER (STEEL) (PSI)	6 X 12 CYLINDER (STEEL) (PSI)	6X6X18 BEAM (STEEL) (PSI)	4 X 8 CYLINDER (CARDBD) (PSI)	4 X 8 CYLINDER (STEEL) (PSI)
CED 0 6/ 8/82 8.5/1.0	BRAND C3 II 813. 15.	NONE 0 0 0	1/2 LIMESTONE BRAND E2 1429. 32. 56.	2.81 BRAND D1 1386. 31. 1.03	NONE 0 -0	.396 .396 323. 4.5 2.0	5.00 148. 77.9 15. MIN DAMP,73F	(28-DAY) • 6970. 6930. 7710. 6260.	(56-DAY) • 8390. 7570. 8750. 8030.	(28-DAY) • 855. 950. 783. 833.	(28-DAY) • 7570. 7930. 7540. 7230.	(28-DAY) • -0• -0 -0 -0
CED 0 6/ 8/82 8.5/1.5	BRAND C3 II 819. 15.	NONE 0 0 0	1/2 LIMESTONE BRAND E2 1730. 39. 67.	2.81 BRAND D1 1117. 25. 1.55	NONE 0 -0	.378 .378 310. 4.3 2.0	3.25 151. 77.9 15. MIN DAMP,73F	(28-DAY) • 7920. 8210. 7890. 7660.	(56-DAY) • 9020. 9160. 8610. 9300.	(28-DAY) • 897. 917. 883. 892.	(28-DAY) • 7830. 7850. 8440. 7210.	(28-DAY) • -0• -0 -0 -0
CED 0 6/10/82 8.5/2.1	BRAND C3 II 788. 15.	NONE 0 0 0	1/2 LIMESTONE BRAND E2 1936. 44. 75.	2.81 BRAND D1 936. 21. 2.07	NONE 0 -0	.394 .394 310. 4.4 2.0	3.00 151. 77.9 15. MIN DAMP,73F	(28-DAY) • 7180. 7370. 6930. 7250.	(56-DAY) • 8490. 8360. 8910. 8210.	(28-DAY) • 913. 0 892. 933.	(28-DAY) • 7700. 7320. 8100. 7670.	(28-DAY) • -0• -0 -0 -0
CED 0 6/10/82 10.0/1.0	BRAND C3 II 923. 17.	NONE 0 0 0	1/2 LIMESTONE BRAND E2 1381. 31. 54.	2.81 BRAND D1 1338. 30. 1.03	NONE 0 -0	.352 .352 324. 4.0 2.0	3.75 150. 78.8 15. MIN DAMP,73F	(28-DAY) • 8170. 8670. 7550. 8290.	(56-DAY) • 9560. 9690. 9580. 9410.	(28-DAY) • 966. 958. 983. 958.	(28-DAY) • 8040. 8660. 8040. 7430.	(28-DAY) • -0• -0 -0 -0
CED 0 6/10/82 10.0/1.5	BRAND C3 II 921. 17.	NONE 0 0 0	1/2 LIMESTONE BRAND E2 1657. 37. 64.	2.81 BRAND D1 1069. 24. 1.55	NONE 0 -0	.351 .351 323. 4.0 2.0	3.25 152. 78.8 15. MIN DAMP,73F	(28-DAY) • 8270. 8840. 7890. 8080.	(56-DAY) • 9060. 8810. 9620. 8750.	(28-DAY) • 972. 1017. 967. 933.	(28-DAY) • 7830. 8040. 7720. 7720.	(28-DAY) • -0• -0 -0 -0
C.D 0 6/14/82 10.0/2.1	BRAND C3 II 920. 17.	NONE 0 0 0	1/2 LIMESTONE BRAND E2 1843. 41. 72.	2.81 BRAND D1 887. 20. 2.08	NONE 0 -0	.351 .351 323. 4.0 2.0	3.00 152. 78.8 15. MIN DAMP,73F	(28-DAY) • 7980. 8190. 7500. 8240.	(56-DAY) • 9400. 9510. 9200. 9500.	(28-DAY) • 920. 917. 950. 892.	(28-DAY) • 6930. 7310. 6600. 6880.	(28-DAY) • -0• -0 -0 -0
CED 0 6/14/82 10.0/1.6	BRAND E1 III 887. 17.	NONE 0 0 0	1/2 LIMESTONE BRAND E2 1598. 36. 62.	2.81 BRAND D1 1626. 23. 1.56	NONE 0 -0	.420 .420 373. 4.7 2.0	3.50 147. 80.6 15. MIN DAMP,73F	(28-DAY) • 7020. 6680. 6840. 7530.	(56-DAY) • 8420. 8380. 8580. 8310.	(28-DAY) • 906. 867. 942. 908.	(28-DAY) • 6800. 7310. 6760. 6340.	(28-DAY) • -0• -0 -0 -0
EED 0 6/15/82 8.5/1.5	BRAND E1 III 773. 15.	NONE 0 0 0	1/2 LIMESTONE BRAND E2 1708. 38. 66.	2.81 BRAND D1 1104. 25. 1.55	NONE 0 -0	.436 .436 337. 4.9 2.0	3.25 148. 80.6 15. MIN DAMP,73F	(28-DAY) • 6900. 6760. 6760. 7180.	(56-DAY) • 8150. 7960. 7960. 8540.	(28-DAY) • 875. 875. 867. 883.	(28-DAY) • 7490. 7490. 7630. 7340.	(28-DAY) • -0• -0 -0 -0

MIX I.D. MIX DATE CF/CAFA	CEMENT BRAND TYPE LBS/CUYD PCT VOLUME	FLYASH BRAND CLASS LBS/CUYD PCT VOLUME PCT REPLACED	COARSE AGG SIZE MATERIAL SOURCE LBS/CUYD P:VOL,DRUW	FINE AGG FINENESS SOURCE LBS/CUYD PCT VOLUME CA/FA(LB/LB)	ADMIXTURE TYPE BRAND DOSE(OZ/100) (2ND TYPE) (2ND DOSE)	WATER W/C W/B LB/CUYD GAL/SACK PCT AIR	MISC SLUMP UNIT WT MIX TMP MX TIME CURING	•••••••••••••••• TEST RESULTS 6 X 12 CYLINDER (STEEL) (PSI)	 6 X 12 CYLINDER (STEEL) (PSI)	 6X6X18 BEAM (STEEL) (PSI)	 4 X 8 CYLINDER (CARDBD) (PSI)	••••••••••••••• 4 X 8 CYLINDER (STEEL) (PSI)
D:D 0	BRAND D1		1/2	2.81		.343	3.50	(28-DAY)	(56-DAY)	(28-DAY)	(28-DAY)	(28-DAY)
6/15/82	I	NONE	LIMESTONE	BRAND D1	NONE	.343	149.	• 7430.	• 8510.	• 1014.	• 7820.	• -0•
10.0/1.5	924.	0	BRAND E2	1076.	0	317.	73.7	7360.	7960.	1083.	7910.	-0
	17.	0	1664.	24.		3.9	15. MIN	7360.	8540.	933.	7980.	-0
		0	37. 65.	1.55	-0	2.0	DAMP,73F	7570.	9020.	1025.	7560.	-0
D:D 0	BRAND D1		1/2	2.81		.402	3.25	(28-DAY)	(56-DAY)	(28-DAY)	(28-DAY)	(28-DAY)
6/16/82	I	NONE	LIMESTONE	BRAND D1	NONE	.402	149.	• 7250.	• 7960.	• 911.	• 7030.	• -0•
8.5/1.6	788.	0	BRAND E2	1115.	0	316.	79.8	6840.	8130.	942.	7190.	-0
	15.	0	1740.	25.		4.5	15. MIN	7300.	8170.	883.	6800.	-0
		0	39. 68.	1.56	-0	2.0	DAMP,73F	7600.	7570.	908.	7100.	-0
CFB 0	BRAND C3		1/2	2.57		.430	3.00	(28-DAY)	(56-DAY)	(28-DAY)	(28-DAY)	(28-DAY)
6/16/82	II	NONE	GRAVEL	BRAND B2	NONE	.430	148.	• 6960.	• 7830.	• 753.	• 7040.	• -0•
7.0/1.0	655.	0	BRAND F1	1477.	0	282.	77.9	7200.	7690.	758.	7230.	-0
	12.	0	1514.	34.		4.9	15. MIN	6450.	7730.	725.	6860.	-0
		0	35. 58.	1.03	-0	2.0	DAMP,73F	7220.	8080.	775.	7020.	-0
CFB 0	BRAND C3		1/2	2.57		.456	3.00	(28-DAY)	(56-DAY)	(28-DAY)	(28-DAY)	(28-DAY)
6/17/82	II	NONE	GRAVEL	BRAND B2	NONE	.456	149.	• 7170.	• 7750.	• 781.	• 5520.	• -0•
7.0/1.5	649.	0	BRAND F1	1166.	0	296.	77.9	7070.	8030.	833.	5430.	-0
	12.	0	1795.	27.		5.1	15. MIN	7320.	7460.	692.	5280.	-0
		0	41. 68.	1.54	-0	2.0	DAMP,73F	7130.	5750.	817.	5860.	-0
CFB 0	BRAND C3		1/2	2.57		.470	4.25	(28-DAY)	(56-DAY)	(28-DAY)	(28-DAY)	(28-DAY)
6/17/82	II	NONE	GRAVEL	BRAND B2	NONE	.470	149.	• 6490.	• 7430.	• 769.	• 5230.	• -0•
7.0/2.0	644.	0	BRAND F1	965.	0	303.	78.8	6370.	7530.	808.	5570.	-0
	12.	0	1981.	22.		5.3	15. MIN	6470.	7340.	750.	5380.	-0
		0	46. 75.	2.05	-0	2.0	DAMP,73F	6370.	7430.	750.	4740.	-0
CFB 0	BRAND C3		1/2	2.57		.370	3.50	(28-DAY)	(56-DAY)	(28-DAY)	(28-DAY)	(28-DAY)
6/18/82	II	NONE	GRAVEL	BRAND B2	NONE	.370	149.	• 7630.	• 8750.	• 872.	• 7840.	• -0•
8.5/1.0	792.	0	BRAND F1	1432.	0	293.	79.7	7660.	8420.	858.	7880.	-0
	15.	0	1417.	33.		4.2	15. MIN	7800.	8950.	817.	8280.	-0
		0	33. 54.	.99	-0	2.0	DAMP,73F	7440.	8890.	942.	7350.	-0
CFB 0	BRAND C3		1/2	2.57		.373	4.25	(28-DAY)	(56-DAY)	(28-DAY)	(28-DAY)	(28-DAY)
6/18/82	II	NONE	GRAVEL	BRAND B2	NONE	.373	150.	• 7710.	• 8820.	• 889.	• 6400.	• -0•
8.5/1.5	792.	0	BRAND F1	1144.	0	295.	77.9	7780.	8720.	900.	6370.	-0
	15.	0	1701.	25.		4.2	15. MIN	7850.	8820.	883.	5960.	-0
		0	39. 65.	1.49	-0	2.0	DAMP,73F	7500.	8930.	883.	6860.	-0
CFB 0	BRAND C3		1/2	2.57		.380	3.50	(28-DAY)	(56-DAY)	(28-DAY)	(28-DAY)	(28-DAY)
7/ 8/82	II	NONE	GRAVEL	BRAND B2	NONE	.380	151.	• 7720.	• 8560.	• 878.	• 7540.	• -0•
8.5/2.0	790.	0	BRAND F1	932.	0	300.	79.7	7500.	8510.	833.	8280.	-0
	15.	0	1902.	22.		4.3	15. MIN	7920.	8740.	883.	7670.	-0
		0	44. 72.	2.04	-0	2.0	DAMP,73F	7730.	8440.	917.	6680.	-0

MIX I.D. MIX DATE CF/CAFA	CEMENT BRAND TYPE LBS/CUYD PCT VOLUME	FLYASH BRAND CLASS LBS/CUYD PCT VOLUME PCT REPLACED	COARSE AGG SIZE MATERIAL SOURCE LBS/CUYD P:VOL,DRUW	FINE AGG FINENESS SOURCE LBS/CUYD PCT VOLUME CA/FA(LB/LB)	ADMIXTURE TYPE BRAND DOSE(OZ/100) (2ND TYPE) (2ND DOSE)	WATER W/C W/B LB/CUYD GAL/SACK PCT AIR	MISC SLUMP UNIT WT MIX TMP MX TIME CURING	TEST RESULTS 6 X 12 CYLINDER (STEEL) (PSI)	6 X 12 CYLINDER (STEEL) (PSI)	6X6X18 BEAM (STEEL) (PSI)	4 X 8 CYLINDER (CARDBD) (PSI)	4 X 8 CYLINDER (STEEL) (PSI)
CFB 0	BRAND C3		1/2	2.57		.371	5.50	(28-DAY)	(56-DAY)	(28-DAY)	(28-DAY)	(28-DAY)
7/ 8/82	II	NONE	GRAVEL	BRAND B2	NONE	.371	151.	• 7770. •	8710. •	931. •	6850. •	-0•
10.0/1.0	908.	0	BRAND F1	1308.	0	337.	78.8	8060.	8720.	942.	7050.	-0
	17.	0	1334.	30.		4.2	15. MIN	7500.	8610.	933.	6650.	-0
		0	31. 51.	1.02	-0	2.0	DAMP,73F	7750.	8810.	917.	6840.	-0
CFB 0	BRAND C3		1/2	2.57		.323	4.25	(28-DAY)	(57-DAY)	(28-DAY)	(28-DAY)	(28-DAY)
7/12/82	II	NONE	GRAVEL	BRAND B2	NONE	.323	148.	• 8490. •	8850. •	958. •	8570. •	-0•
10.0/1.5	929.	0	BRAND F1	1076.	0	300.	78.8	8540.	8470.	900.	8980.	-0
	18.	0	1644.	25.		3.6	15. MIN	8190.	8740.	1017.	8400.	-0
		0	38. 63.	1.53	-0	2.0	DAMP,73F	8740.	9340.	958.	8320.	-0
CFB 0	BRAND C3		1/2	2.57		.320	3.75	(28-DAY)	(57-DAY)	(28-DAY)	(28-DAY)	(28-DAY)
7/12/82	II	NONE	GRAVEL	BRAND B2	NONE	.320	150.	• 8650. •	9030. •	995. •	8270. •	-0•
10.0/2.0	930.	0	BRAND F1	897.	0	297.	79.7	8840.	9020.	1000.	8090.	-0
	18.	0	1830.	21.		3.6	15. MIN	8380.	9390.	992.	8400.	-0
		0	42. 70.	2.04	-0	2.0	DAMP,73F	8740.	8570.	992.	8310.	-0
EFB 0	BRAND E1		1/2	2.57		.390	3.00	(28-DAY)	(56-DAY)	(28-DAY)	(28-DAY)	(28-DAY)
7/15/82	III	NONE	GRAVEL	BRAND B2	NONE	.390	146.	• 8110. •	8520. •	858. •	7190. •	-0•
10.0/1.5	899.	0	BRAND F1	1036.	0	351.	81.5	8380.	8150.	900.	7110.	-0
	17.	0	1578.	24.		4.4	15. MIN	7390.	8670.	858.	7210.	-0
		0	36. 60.	1.52	-0	2.0	DAMP,73F	8560.	8740.	817.	7260.	-0
FFB 0	BRAND E1		1/2	2.57		.442	4.00	(28-DAY)	(56-DAY)	(28-DAY)	(28-DAY)	(28-DAY)
7/15/82	III	NONE	GRAVEL	BRAND B2	NONE	.442	146.	• 7190. •	7929. •	867. •	6770. •	-0•
8.5/1.5	770.	0	BRAND F1	1087.	0	341.	81.5	7060.	8220.	867.	6810.	-0
	15.	0	1659.	25.		5.0	15. MIN	7020.	8030.	858.	7000.	-0
		0	38. 63.	1.53	-0	2.0	DAMP,73F	7480.	7500.	875.	6510.	-0
DFB 0	BRAND D1		1/2	2.57		.367	4.00	(28-DAY)	(56-DAY)	(28-DAY)	(28-DAY)	(28-DAY)
7/13/82	I	NONE	GRAVEL	BRAND B2	NONE	.367	146.	• 7430. •	7590. •	1000. •	7130. •	-0•
10.0/1.5	912.	0	BRAND F1	1043.	0	335.	81.5	7300.	7500.	1050.	6780.	-0
	17.	0	1602.	24.		4.1	15. MIN	7290.	7780.	1017.	7670.	-0
		0	37. 61.	1.54	-0	2.0	DAMP,73F	7710.	7500.	933.	6940.	-0
DFB 0	BRAND D1		1/2	2.57		.402	3.00	(28-DAY)	(56-DAY)	(28-DAY)	(28-DAY)	(28-DAY)
7/13/82	I	NONE	GRAVEL	BRAND B2	NONE	.402	147.	• 7380. •	7760. •	883. •	7460. •	-0•
8.5/1.5	786.	0	BRAND F1	1102.	0	316.	78.8	7990.	7390.	950.	7400.	-0
	15.	0	1695.	25.		4.5	15. MIN	7460.	7570.	908.	7670.	-0
		0	39. 65.	1.54	-0	2.0	DAMP,73F	6700.	8310.	792.	7310.	-0
CFC A	BRAND C4		1/2	2.72	SUPERPLSTCZR	.335	4.50	(28-DAY)	(56-DAY)	(28-DAY)	(28-DAY)	(28-DAY)
8/13/82	II	NONE	LIMESTONE	BRAND C1	BRAND A1	.335	153.	• 9530. •	9640. •	930. •	9340. •	-0•
7.0/1.0	651.	0	BRAND E2	1606.	15.0	218.	77.9	9340.	9140.	983.	9340.	-0
(W)	12.	0	1620.	36.		3.8	15. MIN	9500.	5450.	875.	9440.	-0
		0	36. 63.	1.01	-0	2.0	DAMP,73F	9740.	10130.	933.	9250.	-0

	CEMENT	FLYASH	COARSE AGG	FINE AGG	ADMIXTURE	WATER	MISC	•••••••••••••• TEST	RESULTS	••••••••••••••		
MIX I.D. MIX DATE CF/CAFA	BRAND TYPE LBS/CUYD PCT VOLUME	BRAND CLASS LBS/CUYD PCT VOLUME PCT REPLACED	SIZE MATERIAL SOURCE LBS/CUYD P:VOL,DRUM	FINENESS SOURCE LBS/CUYD PCT VOLUME CA/FA(LB/LB)	TYPE BRAND DOSE(OZ/100) (2ND TYPE) (2ND DOSE)	W/C W/H LB/CUYD GAL/SACK PCT AIR	SLUMP UNIT WT MIX TMP MX TIME CURING	6 X 12 CYLINDER (STEEL) (PSI)	6 X 12 CYLINDER (STEEL) (PSI)	6X6X18 BEAM (STEEL) (PSI)	4 X 8 CYLINDER (CARDBD) (PSI)	4 X 8 CYLINDER (STEEL) (PSI)
CIC A	BRAND C3		1/2	2.72	SUPERPLSTCZR	.409	2.88	(29-DAY)	(56-DAY)	(29-DAY)	(29-DAY)	(29-DAY)
8/ 9/82	II	NONE	LIMESTONE	BRAND C1	BRAND A1	.409	155.	• 8270.	• 9090.	• 831.	• 7240.	• -0•
7.0/1.5	631.	0	BRAND E2	1253.	15.0	258.	80.6	8360.	9040.	858.	7260.	-0
	12.	0	1888.	28.		4.6	15. MIN	8130.	9070.	817.	6850.	-0
		0	42. 73.	1.51	-0	2.0	DAMP,73F	8330.	9160.	818.	7210.	-0
CIC A	BRAND C4		1/2	2.72	SUPERPLSTCZR	.376	12.00	(28-DAY)	(56-DAY)	(28-DAY)	(28-DAY)	(28-DAY)
8/12/82	II	NONE	LIMESTONE	BRAND C1	BRAND A1	.376	155.	• 8710.	• 9020.	• 906.	• 7970.	• -0•
7.0/2.0	639.	0	BRAND E2	1058.	15.0	241.	77.0	8910.	7800.	925.	7670.	-0
	12.	0	2122.	24.		4.2	15. MIN	8650.	8580.	907.	7800.	-0
		0	48. 53.	2.01	-0	2.0	DAMP,73F	8580.	9460.	887.	8440.	-0
CIC A	BRAND C3		1/2	2.72	SUPERPLSTCZR	.380	3.75	(29-DAY)	(56-DAY)	(29-DAY)	(29-DAY)	(29-DAY)
8/ 9/82	II	NONE	LIMESTONE	BRAND C1	BRAND A1	.380	155.	• 8910.	• 9470.	• 959.	• 8150.	• -0•
8.5/1.0	799.	0	BRAND E2	1521.	10.3	240.	77.9	8820.	9250.	1038.	8860.	-0
	15.	0	1524.	34.		3.4	••• MIN	9650.	9510.	983.	7290.	-0
		0	34. 59.	1.00	-0	2.0	DAMP,73F	8260.	9640.	857.	8310.	-0
CFC A	BRAND C3		1/2	2.72	SUPERPLSTCZR	.305	3.25	(28-DAY)	(56-DAY)	(28-DAY)	(28-DAY)	(28-DAY)
8/ 5/82	II	NONE	LIMESTONE	BRAND C1	BRAND A1	.305	155.	• 9640.	• 10220.	• 1069.	• 9170.	• -0•
8.5/1.5	794.	0	BRAND E2	1228.	10.1	243.	78.8	9550.	10190.	1042.	9550.	-0
	15.	0	1816.	28.		3.4	15. MIN	10560.	10190.	1133.	8910.	-0
		0	41. 71.	1.48	-0	2.0	DAMP,73F	8810.	10270.	1033.	9060.	-0
CIC A	BRAND C3		1/2	2.72	SUPERPLSTCZR	.305	3.25	(28-DAY)	(56-DAY)	(28-DAY)	(28-DAY)	(28-DAY)
8/10/82	II	NONE	LIMESTONE	BRAND C1	BRAND A1	.305	154.	• 8920.	• 9670.	• 1010.	• 8020.	• -0•
8.5/2.0	797.	0	BRAND E2	1012.	8.9	243.	78.8	7780.	9640.	1092.	8120.	-0
	15.	0	2030.	23.		3.4	15. MIN	9370.	9120.	910.	8370.	-0
		0	46. 79.	2.01	-0	2.0	DAMP,73F	9600.	10260.	1028.	7580.	-0
CIC A	BRAND C3		1/2	2.72	SUPERPLSTCZR	.298	5.25	(28-DAY)	(56-DAY)	(28-DAY)	(28-DAY)	(28-DAY)
8/10/82	II	NONE	LIMESTONE	BRAND C1	BRAND A1	.298	153.	• 9020.	• 9990.	• 1120.	• 8100.	• -0•
10.0/1.0	941.	0	BRAND E2	1407.	8.3	281.	73.7	8740.	10030.	1102.	8010.	-0
	18.	0	1412.	32.		3.4	15. MIN	10010.	10170.	983.	9020.	-0
		0	32. 55.	1.00	-0	2.0	DAMP,73F	8310.	9780.	1300.	7260.	-0
CEC A	BRAND C4		1/2	2.72	SUPERPLSTCZR	.299	7.00	(28-DAY)	(56-DAY)	(28-DAY)	(28-DAY)	(28-DAY)
8/12/82	II	NONE	LIMESTONE	BRAND C1	BRAND A1	.299	151.	• 8360.	• 9250.	• 992.	• 7480.	• -0•
10.0/1.5	941.	0	BRAND E2	1127.	6.0	281.	79.7	8210.	9510.	1017.	8280.	-0
	18.	0	1693.	26.		3.4	15. MIN	6370.	8980.	945.	7160.	-0
		0	38. 66.	1.50	-0	2.0	DAMP,73F	8510.	7550.	1013.	7000.	-0
CIC A	BRAND C3		1/2	2.72	SUPERPLSTCZR	.287	5.00	(28-DAY)	(56-DAY)	(28-DAY)	(28-DAY)	(28-DAY)
8/11/82	II	NONE	LIMESTONE	BRAND C1	BRAND A1	.287	153.	• 8970.	• 9490.	• 1042.	• 8230.	• -0•
10.0/2.0	946.	0	BRAND E2	943.	7.1	271.	78.8	8770.	9710.	1018.	8720.	-0
	18.	0	1899.	21.		3.2	15. MIN	9110.	9050.	1050.	8410.	-0
		0	43. 74.	2.01	-0	2.0	DAMP,73F	9040.	9710.	1057.	7570.	-0

MIX I.D. MIX DATE CF/CAFA	CEMENT BRAND TYPE LBS/CUYD PCT VOLUME	FLYASH BRAND CLASS LBS/CUYD PCT VOLUME PCT REPLACED	COARSE AGG SIZE MATERIAL SOURCE LBS/CUYD P:VOL,DRUW	FINE AGG FINENESS SOURCE LBS/CUYD PCT VOLUME CA/FA(LB/LB)	ADMIXTURE TYPE BRAND DOSE(OZ/100) (2ND TYPE) (2ND DOSE)	WATER W/C W/B LB/CUYD GAL/SACK PCT AIR	MISC SLUMP UNIT WT MIX TMP MX TIME CURING	TEST RESULTS 6 X 12 CYLINDER (STEEL) (PSI)	6 X 12 CYLINDER (STEEL) (PSI)	6X6X18 BEAM (STEEL) (PSI)	4 X 8 CYLINDER (CAPDBD) (PSI)	4 X 8 CYLINDER (STEEL) (PSI)
CIC A 8/ 4/82 10.0/1.5	BRAND C3 II 938. 18.	NONE 0 0 0	1/2 LIMESTONE BRAND E2 1692. 38. 66.	2.72 BRAND C1 1121. 25. 1.51	SUPERPLSTCZR BRAND A1 15.0 -0	.303 .303 285. 3.4 2.0	12.00 152. 79.7 15. MIN DAMP,73F	(28-DAY) • 9690. 9670. 9850. 9550.	(56-DAY) • 10590. 9540. 10860. 10310.	(29-DAY) • 1004. 1050. 958. 0	(28-DAY) • 8240. 8750. 7290. 8690.	(28-DAY) • -0• -0 -0 -0
CIC A 8/ 3/82 7.0/1.0 (W)	BRAND C3 II 650. 12.	NONE 0 0 0	1/2 LIMESTONE BRAND E2 1619. 36. 63.	2.72 BRAND C1 1622. 37. 1.00	SUPERPLSTCZR BRAND A1 25.0 -0	.327 .327 212. 3.7 2.0	4.00 155. 79.7 15. MIN DAMP,73F	(-0-DAY) • -0 -0 -0 -0	(56-DAY) • 6490. 6610. 6310. 6540.	(-0-DAY) • -0 -0 -0 -0	(-0-DAY) • -0 -0 -0 -0	(-0-DAY) • -0• -0 -0 -0
CIC A 8/11/82 7.0/1.0 (W)	BRAND C3 II 637. 12.	NONE 0 0 0	1/2 LIMESTONE BRAND E2 1590. 36. 62.	2.72 BRAND C1 1583. 36. 1.00	SUPERPLSTCZR BRAND A1 23.9 -0	.380 .380 242. 4.3 2.0	12.00 149. 77.0 15. MIN DAMP,73F	(28-DAY) • 5870. 5660. 5750. 6210.	(56-DAY) • 6300. 6120. 6490. 6280.	(28-DAY) • 672. 693. 643. 680.	(28-DAY) • 7130. 6930. 5790. 7320.	(28-DAY) • -0• -0 -0 -0
CIC B 8/17/82 7.0/1.0 (P)	BRAND C4 II 641. 12.	NONE 0 0 0	1/2 LIMESTONE BRAND E2 1596. 36. 62.	2.72 BRAND C1 1523. 35. 1.05	SUPERPLSTCZR BRAND B1 8.2 -0	.408 .408 262. 4.6 2.0	5.25 151. 77.9 15. MIN DAMP,73F	(28-DAY) • 7320. 7220. 7600. 7130.	(56-DAY) • 7140. 6950. 6950. 7530.	(28-DAY) • 822. 800. 850. 817.	(28-DAY) • 6380. 5790. 6570. 6780.	(28-DAY) • -0• -0 -0 -0
CIC B 8/17/82 7.0/1.6	BRAND C4 II 647. 12.	NONE 0 0 0	1/2 LIMESTONE BRAND E2 1935. 43. 75.	2.72 BRAND C1 1230. 28. 1.57	SUPERPLSTCZR BRAND B1 8.8 -0	.376 .376 244. 4.2 2.0	4.50 152. 80.6 15. MIN DAMP,73F	(28-DAY) • 7610. 8030. 7520. 7290.	(56-DAY) • 7780. 8710. 7550. 7780.	(29-DAY) • 839. 858. 833. 825.	(28-DAY) • 6980. 7240. 6460. 7240.	(28-DAY) • -0• -0 -0 -0
CIC B 8/18/82 7.0/2.0	BRAND C4 II 647. 12.	NONE 0 0 0	1/2 LIMESTONE BRAND E2 2152. 48. 84.	2.72 BRAND C1 1062. 24. 2.03	SUPERPLSTCZR BRAND B1 15.0 -0	.348 .348 225. 3.9 2.0	12.00 156. 77.9 15. MIN DAMP,73F	(28-DAY) • 9210. 8970. 5580. 9440.	(56-DAY) • 9510. 9210. 7570. 9800.	(28-DAY) • 929. 917. 967. 900.	(28-DAY) • 8420. 9260. 8310. 7690.	(28-DAY) • -0• -0 -0 -0
CIC B 8/18/82 8.5/1.0 (N)	BRAND C4 II 795. 15.	NONE 0 0 0	1/2 LIMESTONE BRAND E2 1517. 34. 59.	2.72 BRAND C1 1501. 34. 1.01	SUPERPLSTCZR BRAND B1 6.0 -0	.316 .316 251. 3.6 2.0	5.25 151. 78.8 15. MIN DAMP,73F	(28-DAY) • 7740. 7850. 8010. 7370.	(56-DAY) • 8570. 8280. 8610. 8810.	(29-DAY) • 925. 925. 917. 933.	(28-DAY) • 7940. 7510. 7640. 8660.	(28-DAY) • -0• -0 -0 -0
CIC B 8/19/82 8.5/1.5 (L)	BRAND C4 II 789. 15.	NONE 0 0 0	1/2 LIMESTONE BRAND E2 1806. 41. 70.	2.72 BRAND C1 1179. 27. 1.53	SUPERPLSTCZR BRAND H1 7.4 -0	.338 .338 267. 3.8 2.0	4.75 154. 77.9 15. MIN DAMP,73F	(28-DAY) • 8320. 8700. 8380. 7870.	(56-DAY) • 8860. 8840. 8980. 8750.	(28-DAY) • 930. 952. 970. 868.	(28-DAY) • 7620. 7290. 7970. 7610.	(28-DAY) • -0• -0 -0 -0

	CEMENT	FLYASH	COARSE AGG	FINE AGG	ADMIXTURE	WATER	MISC	•••••••••••••• TEST RESULTS	••••••••••••••			
MIX I.D. MIX DATE CF/CAFA	BRAND TYPE LBS/CUYD PCT VOLUME	BRAND CLASS LBS/CUYD PCT VOLUME PCT REPLACED	SIZE MATERIAL SOURCE LBS/CUYD P:VOL,DRUW	FINENESS SOURCE LBS/CUYD PCT VOLUME CA/FA(LB/LB)	TYPE BRAND DOSE(OZ/100) (2ND TYPE) (2ND DOSE)	W/C W/B LB/CUYD GAL/SACK PCT AIR	SLUMP UNIT WT MIX TMP MX TIME CURING	6 X 12 CYLINDER (STEEL) (PSI)	6 X 12 CYLINDER (STEEL) (PSI)	6X6X18 BEAM (STEEL) (PSI)	4 X 8 CYLINDER (CAPDRO) (PSI)	4 X 8 CYLINDER (STEEL) (PSI)
CIC B	BRAND C4		1/2	2.72	SUPERPLSTCZR	.334	4.25	(28-DAY)	(56-DAY)	(28-DAY)	(28-DAY)	(28-DAY)
8/19/82	II	NONE	LIMESTONE	BRAND C1	BRAND B1	.334	155.	8610.	9130.	950.	7490.	-0.
8.5/2.0	790.	0	BRAND E2	986.	7.6	264.	77.9	8080.	8980.	947.	7640.	-0
(X)	15.	0	2007.	22.		3.8	15. MIN	8790.	8280.	982.	7460.	-0
		0	45. 78.	2.04	-0	2.0	DAMP,73F	8950.	9270.	922.	7370.	-0
CIC B	BRAND C4		1/2	2.72	SUPERPLSTCZR	.310	5.00	(28-DAY)	(56-DAY)	(28-DAY)	(28-DAY)	(28-DAY)
8/20/82	II	NONE	LIMESTONE	BRAND C1	BRAND B1	.310	150.	8770.	8770.	1011.	7710.	-0.
10.0/1.0	937.	0	BRAND E2	1390.	6.0	290.	78.8	8770.	8910.	1033.	7530.	-0
	18.	0	1407.	31.		3.5	15. MIN	8820.	8720.	1000.	8280.	-0
		0	32. 55.	1.01	-0	2.0	DAMP,73F	8720.	8670.	1000.	7310.	-0
CIC B	BRAND C4		1/2	2.72	SUPERPLSTCZR	.315	6.25	(28-DAY)	(56-DAY)	(28-DAY)	(28-DAY)	(28-DAY)
8/20/82	II	NONE	LIMESTONE	BRAND C1	BRAND B1	.315	153.	8500.	9040.	922.	7640.	-0.
10.0/1.5	934.	0	BRAND E2	1109.	6.0	294.	78.8	8220.	9140.	900.	7080.	-0
(Y)	18.	0	1683.	25.		3.5	15. MIN	8450.	8930.	992.	8000.	-0
		0	38. 65.	1.52	-0	2.0	DAMP,73F	8840.	7850.	875.	7850.	-0
CIC B	BRAND C4		1/2	2.45	SUPERPLSTCZR	.380	4.00	(28-DAY)	(56-DAY)	(28-DAY)	(28-DAY)	(28-DAY)
9/14/82	II	NONE	LIMESTONE	BRAND C2	BRAND B1	.380	150.	7580.	8830.	839.	8070.	-0.
7.0/1.0	660.	0	BRAND E3	1580.	12.1	251.	77.9	7660.	8770.	833.	7930.	-0
(P)	12.	0	1563.	36.		4.3	15. MIN	8120.	9120.	867.	8320.	-0
		0	35. 62.	.99	-0	2.0	DAMP,73F	6950.	8590.	817.	7960.	-0
CIC B	BRAND C4		1/2	2.45	SUPERPLSTCZR	.283	4.50	(28-DAY)	(56-DAY)	(28-DAY)	(28-DAY)	(28-DAY)
9/21/82	II	NONE	LIMESTONE	BRAND C2	BRAND B1	.283	153.	9580.	9970.	1056.	9630.	-0.
8.5/1.0	809.	0	BRAND E3	1536.	14.4	229.	78.8	9430.	10060.	1025.	10060.	-0
(N)	15.	0	1542.	35.		3.2	15. MIN	9800.	9540.	1072.	9490.	-0
		0	35. 61.	1.00	-0	2.0	DAMP,73F	9510.	10220.	1071.	9340.	-0
CIC B	BRAND C4		1/2	2.45	SUPERPLSTCZR	.300	4.75	(28-DAY)	(56-DAY)	(28-DAY)	(28-DAY)	(28-DAY)
9/22/82	II	NONE	LIMESTONE	BRAND C2	BRAND B1	.300	153.	9070.	9680.	997.	8020.	-0.
8.5/1.5	803.	0	BRAND E3	1223.	11.0	241.	78.8	8930.	9340.	1025.	7460.	-0
(L)	15.	0	1827.	27.		3.4	15. MIN	9040.	9870.	1025.	8180.	-0
		0	41. 73.	1.49	-0	2.0	DAMP,73F	9250.	9830.	942.	8420.	-0
CIC B	BRAND C4		1/2	2.45	SUPERPLSTCZR	.298	5.00	(28-DAY)	(56-DAY)	(28-DAY)	(28-DAY)	(28-DAY)
9/23/82	II	NONE	LIMESTONE	BRAND C2	BRAND B1	.298	156.	9420.	10380.	1005.	8940.	-0.
8.5/2.0	803.	0	BRAND E3	1008.	13.3	240.	78.8	9250.	10360.	1008.	8390.	-0
(X)	15.	0	2045.	23.		3.4	15. MIN	9200.	10270.	950.	8930.	-0
		0	46. 81.	2.03	-0	2.0	DAMP,73F	9810.	10520.	1058.	9490.	-0
CIC B	BRAND C4		1/2	2.45	SUPERPLSTCZR	.291	5.75	(28-DAY)	(56-DAY)	(28-DAY)	(28-DAY)	(28-DAY)
9/23/82	II	NONE	LIMESTONE	BRAND C2	BRAND B1	.291	154.	8860.	8950.	1083.	8660.	-0.
10.0/1.5	949.	0	BRAND E3	1124.	9.6	277.	79.7	8280.	8930.	1083.	8790.	-0
(Y)	18.	0	1709.	25.		3.3	15. MIN	9090.	8860.	1058.	8340.	-0
		0	38. 68.	1.52	-0	2.0	DAMP,73F	9200.	9070.	1108.	8860.	-0

MIX I.D. / MIX DATE / CF/CAFA	CEMENT: BRAND / TYPE / LBS/CUYD / PCT VOLUME	FLYASH: BRAND / CLASS / LBS/CUYD / PCT VOLUME / PCT REPLACED	COARSE AGG: SIZE / MATERIAL / SOURCE / LBS/CUYD / P:VOL,DRUW	FINE AGG: FINENESS / SOURCE / LBS/CUYD / PCT VOLUME / CA/FA(LB/LB)	ADMIXTURE: TYPE / BRAND / DOSE(OZ/100) / (2ND TYPE) / (2ND DOSE)	WATER: W/C / W/B / LB/CUYD / GAL/SACK / PCT AIR	MISC: SLUMP / UNIT WT / MIX TMP / MX TIME / CURING	TEST RESULTS: 6 X 12 CYLINDER (STEEL) (PSI)	6 X 12 CYLINDER (STEEL) (PSI)	6X6X18 BEAM (STEEL) (PSI)	4 X 8 CYLINDER (CARDBD) (PSI)	4 X 8 CYLINDER (STEEL) (PSI)
CEC B	BRAND C4		1/2	2.45	SUPERPLSTCZR	.289	5.00	(28-DAY)	(56-DAY)	(28-DAY)	(28-DAY)	(28-DAY)
9/27/82	II	NONE	LIMESTONE	BRAND C2	BRAND B1	.289	155.	• 8780.	• 9150.	• 1017.	• 8160.	• -0•
10.0/2.0	949.	0	BRAND E3	950.	8.6	274.	78.8	9090.	8820.	1067.	8130.	-0
	18.	0	1892.	21.		3.3	15. MIN	8750.	9300.	975.	8280.	-0
		0	43. 75.	1.99	-0	2.0	DAMP,73F	8490.	9340.	1008.	8070.	-0
CFB B	BRAND C4		1/2	2.57	SUPERPLSTCZR	.333	4.25	(28-DAY)	(56-DAY)	(28-DAY)	(28-DAY)	(28-DAY)
8/24/82	II	NONE	GRAVEL	BRAND B2	BRAND B1	.333	149.	• 8290.	• 8470.	• 859.	• 7600.	• -0•
7.0/1.0	652.	0	BRAND F1	1570.	15.0	217.	78.8	8210.	8350.	858.	7500.	-0
	12.	0	1589.	36.		3.7	15. MIN	8290.	8490.	862.	7290.	-0
		0	37. 60.	1.01	-0	2.0	DAMP,73F	8380.	8580.	858.	8010.	-0
CFB B	BRAND C4		1/2	2.57	SUPERPLSTCZR	.334	5.25	(28-DAY)	(56-DAY)	(28-DAY)	(28-DAY)	(28-DAY)
8/24/82	II	NONE	GRAVEL	BRAND B2	BRAND B1	.334	151.	• 8970.	• 9080.	• 847.	• 7460.	• -0•
7.0/1.5	652.	0	BRAND F1	1254.	15.0	217.	78.8	9090.	9480.	833.	7640.	-0
	12.	0	1906.	29.		3.7	15. MIN	8840.	9090.	842.	7130.	-0
		0	44. 73.	1.52	-0	2.0	DAMP,73F	7570.	8670.	867.	7610.	-0
CFB B	BRAND C4		1/2	2.57	SUPERPLSTCZR	.351	4.00	(28-DAY)	(56-DAY)	(28-DAY)	(28-DAY)	(28-DAY)
8/25/82	II	NONE	GRAVEL	BRAND B2	BRAND B1	.351	152.	• 8920.	• 9450.	• 921.	• 8820.	• -0•
7.0/2.0	645.	0	BRAND F1	1047.	15.0	226.	79.7	9120.	8950.	848.	7960.	-0
	12.	0	2098.	24.		3.9	15. MIN	8770.	9690.	923.	8440.	-0
		0	48. 80.	2.00	-0	2.0	DAMP,73F	8860.	9710.	993.	7660.	-0
CFB B	BRAND C4		1/2	2.57	SUPERPLSTCZR	.290	5.50	(28-DAY)	(56-DAY)	(29-DAY)	(28-DAY)	(28-DAY)
8/25/82	II	NONE	GRAVEL	BRAND B2	BRAND B1	.290	151.	• 8660.	• 9230.	• 1004.	• 8040.	• -0•
8.5/1.0	802.	0	BRAND F1	1498.	10.6	233.	78.8	8810.	9530.	1013.	8260.	-0
	15.	0	1499.	35.		3.3	15. MIN	8470.	8970.	982.	7940.	-0
		0	34. 57.	1.00	-0	2.0	DAMP,73F	8700.	9200.	1017.	7930.	-0
CFB B	BRAND C4		1/2	2.57	SUPERPLSTCZR	.300	4.25	(28-DAY)	(56-DAY)	(29-DAY)	(28-DAY)	(28-DAY)
8/26/82	II	NONE	GRAVEL	BRAND B2	BRAND B1	.300	151.	• 8370.	• 9160.	• 1061.	• 7170.	• -0•
8.5/1.5	801.	0	BRAND F1	1192.	9.0	240.	78.8	7920.	9050.	1025.	7290.	-0
	15.	0	1788.	28.		3.4	15. MIN	8360.	9580.	1092.	7170.	-0
		0	41. 68.	1.50	-0	2.0	DAMP,73F	8840.	8840.	1067.	7030.	-0
CFB B	BRAND C4		1/2	2.57	SUPERPLSTCZR	.307	5.50	(28-DAY)	(56-DAY)	(28-DAY)	(28-DAY)	(28-DAY)
8/26/82	II	NONE	GRAVEL	BRAND B2	BRAND B1	.307	152.	• 8370.	• 9200.	• 974.	• 7000.	• -0•
8.5/2.0	798.	0	BRAND F1	986.	10.1	245.	79.7	8260.	9320.	975.	6560.	-0
	15.	0	1983.	23.		3.5	15. MIN	8560.	8180.	1038.	7450.	-0
		0	46. 75.	2.01	-0	2.0	DAMP,73F	8290.	9900.	908.	7000.	-0
CFB B	BRAND C4		1/2	2.57	SUPERPLSTCZR	.293	5.00	(28-DAY)	(56-DAY)	(28-DAY)	(28-DAY)	(28-DAY)
8/27/82	II	NONE	GRAVEL	BRAND B2	BRAND B1	.293	150.	• 8930.	• 9320.	• 1050.	• 8170.	• -0•
10.0/1.0	945.	0	BRAND F1	1383.	7.4	277.	80.6	8590.	8720.	1080.	8350.	-0
	18.	0	1384.	32.		3.3	15. MIN	9120.	9370.	1013.	7720.	-0
		0	32. 53.	1.00	-0	2.0	DAMP,73F	9070.	9870.	1057.	8430.	-0

MIX I.D. / MIX DATE / CF/CAFA	CEMENT: BRAND / TYPE / LBS/CUYD / PCT VOLUME	FLYASH: BRAND / CLASS / LBS/CUYD / PCT VOLUME / PCT REPLACED	COARSE AGG: SIZE / MATERIAL / SOURCE / LBS/CUYD / P:VOL,DRUW	FINE AGG: FINENESS / SOURCE / LBS/CUYD / PCT VOLUME / CA/FA(LB/LB)	ADMIXTURE: TYPE / BRAND / DOSE(OZ/100) / (2ND TYPE) / (2ND DOSE)	WATER: W/C / W/B / LB/CUYD / GAL/SACK / PCT AIR	MISC: SLUMP / UNIT WT / MIX TMP / MX TIME / CURING	TEST RESULTS: 6 X 12 CYLINDER (STEEL) (PSI)	6 X 12 CYLINDER (STEEL) (PSI)	6X6X18 BEAM (STEEL) (PSI)	4 X 8 CYLINDER (CARDBD) (PSI)	4 X 8 CYLINDER (STEEL) (PSI)
CFB B	BRAND C4		1/2	2.57	SUPERPLSTCZR	.302	5.50	(28-DAY)	(56-DAY)	(28-DAY)	(28-DAY)	(28-DAY)
8/27/82	II	NONE	GRAVEL	BRAND B2	BRAND B1	.302	151.	• 9070. •	9420. •	1039. •	7660. •	-0•
10.0/1.5	940.	0	BRAND F1	1100.	6.0	284.	80.6	8970.	9390.	1046.	8240.	-0
	18.	0	1653.	25.		3.4	15. MIN	9180.	9440.	1013.	7160.	-0
		0	38. 63.	1.50	-0	2.0	DAMP,73F	9050.	9430.	1057.	7590.	-0
CFB B	BRAND C5		1/2	2.85	SUPERPLSTCZR	.299	5.00	(28-DAY)	(56-DAY)	(28-DAY)	(28-DAY)	(28-DAY)
10/25/82	II	NONE	GRAVEL	BRAND B3	BRAND B1	.299	151.	• 8340. •	8930. •	939. •	7450. •	-0•
10.0/2.0	943.	0	BRAND F2	921.	6.9	282.	-0	8130.	8740.	975.	7580.	-0
	18.	0	1835.	21.		3.4	15. MIN	8260.	8820.	858.	7400.	-0
		0	42. 70.	1.99	-0	2.0	DAMP,73F	8630.	8930.	983.	7360.	-0
CEB B	BRAND C4		1/2	2.57	SUPERPLSTCZR	.340	7.00	(28-DAY)	(56-DAY)	(28-DAY)	(28-DAY)	(28-DAY)
9/28/82	II	NONE	LIMESTONE	BRAND B2	BRAND B1	.340	148.	• 8220. •	8740. •	956. •	7660. •	-0•
7.0/1.0	647.	0	BRAND E3	1604.	15.0	220.	77.9	8310.	8880.	850.	7670.	-0
	12.	0	1588.	37.		3.8	15. MIN	8100.	8750.	850.	7670.	-0
		0	36. 63.	.99	-0	2.0	DAMP,73F	8240.	8580.	867.	7640.	-0
CEB B	BRAND C4		1/2	2.57	SUPERPLSTCZR	.347	4.75	(28-DAY)	(56-DAY)	(28-DAY)	(28-DAY)	(28-DAY)
9/29/82	II	NONE	LIMESTONE	BRAND B2	BRAND B1	.347	156.	• 8770. •	9520. •	926. •	8440. •	-0•
7.0/1.5	647.	0	BRAND E3	1271.	15.0	225.	77.9	8420.	8980.	922.	8230.	-0
	12.	0	1918.	29.		3.9	15. MIN	8900.	9600.	975.	8730.	-0
		0	43. 76.	1.51	-0	2.0	DAMP,73F	8980.	9970.	880.	8370.	-0
CEB B	BRAND C4		1/2	2.57	SUPERPLSTCZR	.349	5.00	(28-DAY)	(60-DAY)	(28-DAY)	(28-DAY)	(28-DAY)
9/30/82	II	NONE	LIMESTONE	BRAND B2	BRAND B1	.349	154.	• 8670. •	9460. •	914. •	8510. •	-0•
7.0/2.0	645.	0	BRAND E3	1072.	15.0	225.	78.8	8740.	7960.	950.	8210.	-0
	12.	0	2123.	25.		3.9	15. MIN	8670.	9570.	933.	8820.	-0
		0	48. 84.	1.98	-0	2.0	DAMP,73F	8610.	9350.	858.	8510.	-0
CEB B	BRAND C4		1/2	2.57	SUPERPLSTCZR	.284	4.75	(28-DAY)	(60-DAY)	(29-DAY)	(28-DAY)	(28-DAY)
9/30/82	II	NONE	LIMESTONE	BRAND B2	BRAND B1	.284	152.	• 9450. •	9950. •	1061. •	8460. •	-0•
8.5/1.0	804.	0	BRAND E3	1527.	12.3	228.	78.8	9580.	9710.	1108.	8790.	-0
	15.	0	1514.	35.		3.2	15. MIN	9760.	10420.	1008.	8420.	-0
		0	34. 60.	.99	-0	2.0	DAMP,73F	9020.	9730.	1067.	8160.	-0
CEB B	BRAND C4		1/2	2.57	SUPERPLSTCZR	.303	4.75	(28-DAY)	(56-DAY)	(28-DAY)	(28-DAY)	(28-DAY)
10/ 4/82	II	NONE	LIMESTONE	BRAND B2	BRAND B1	.303	157.	• 8720. •	9460. •	968. •	7920. •	-0•
8.5/1.5	801.	0	BRAND E3	1193.	10.2	243.	79.7	8840.	9690.	968.	7660.	-0
	15.	0	1821.	28.		3.4	15. MIN	8490.	8970.	942.	8130.	-0
		0	41. 72.	1.53	-0	2.0	DAMP,73F	8840.	9710.	993.	7960.	-0
CEB B	BRAND C4		1/2	2.57	SUPERPLSTCZR	.273	4.75	(28-DAY)	(56-DAY)	(28-DAY)	(28-DAY)	(28-DAY)
10/ 5/82	II	NONE	LIMESTONE	BRAND B2	BRAND B1	.273	156.	• 9850. •	10100. •	1128. •	10130. •	-0•
8.5/2.0	807.	0	BRAND E3	1032.	14.8	220.	78.8	9870.	9410.	1133.	9870.	-0
(R)	15.	0	2041.	24.		3.1	15. MIN	10150.	10360.	1133.	10270.	-0
		0	46. 81.	1.98	-0	2.0	DAMP,73F	9530.	10520.	1117.	10250.	-0

MIX I.D. MIX DATE CF/CAFA	CEMENT BRAND TYPE LBS/CUYD PCT VOLUME	FLYASH BRAND CLASS LBS/CUYD PCT VOLUME PCT REPLACED	COARSE AGG SIZE MATERIAL SOURCE LBS/CUYD P:VOL,DRUW	FINE AGG FINENESS SOURCE LBS/CUYD PCT VOLUME CA/FA(LB/LB)	ADMIXTURE TYPE BRAND DOSE(OZ/100) (2ND TYPE) (2ND DOSE)	WATER W/C W/B LB/CUYD GAL/SACK PCT AIR	MISC SLUMP UNIT WT MIX TMP MX TIME CURING	••••••••••••••• TEST RESULTS 6 X 12 CYLINDER (STEEL) (PSI)	 6 X 12 CYLINDER (STEEL) (PSI)	 6X6X18 BEAM (STEEL) (PSI)	 4 X 8 CYLINDER (CARDBD) (PSI)	••••••••••••••• 4 X 8 CYLINDER (STEEL) (PSI)
CBB B 10/ 6/82 10.3/1.0	BRAND C4 II 943. 18.	 NONE 0 0 0	1/2 LIMESTONE BRAND E3 1396. 31. 55.	2.57 BRAND B2 1402. 32. 1.00	SUPERPLSTCZR BRAND B1 8.9 -0	.295 .295 278. 3.3 2.0	7.00 147. 80.6 15. MIN DAMP,73F	(28-DAY) • 8310. 8190. 7830. 8910.	(56-DAY) • 9270. 9230. 9070. 9500.	(28-DAY) • 978. 1008. 977. 950.	(28-DAY) • 7620. 6960. 7590. 8310.	(28-DAY) • -0• -0 -0 -0
CBB B 10/ 7/82 10.0/1.5	BRAND C5 II 945. 18.	 NONE 0 0 0	1/2 LIMESTONE BRAND E3 1687. 38. 67.	2.57 BRAND B2 1126. 26. 1.50	SUPERPLSTCZR BRAND B1 7.7 -0	.290 .290 274. 3.3 2.0	4.00 152. 80.6 15. MIN DAMP,73F	(28-DAY) • 8920. 9090. 9200. 8470.	(56-DAY) • 9270. 8910. 9230. 9780.	(28-DAY) • 934. 867. 992. 942.	(28-DAY) • 7790. 7990. 7800. 7590.	(28-DAY) • -0• -0 -0 -0
CBB B 10/ 7/82 10.0/2.0	BRAND C5 II 944. 18.	 NONE 0 0 0	1/2 LIMESTONE BRAND E3 1876. 42. 74.	2.57 BRAND B2 939. 22. 2.00	SUPERPLSTCZR BRAND B1 8.0 -0	.292 .292 275. 3.3 2.0	4.25 152. 80.6 15. MIN DAMP,73F	(28-DAY) • 8870. 9040. 8740. 8840.	(56-DAY) • 9290. 8820. 9350. 9510.	(28-DAY) • 970. 1017. 992. 900.	(28-DAY) • 8210. 7240. 8860. 8530.	(28-DAY) • -0• -0 -0 -0
CBB B 10/21/82 7.0/1.0	BRAND C5 II 653. 12.	 NONE 0 0 0	3/4 LIMESTONE BRAND B2 1580. 36. 61.	2.85 BRAND B3 1581. 37. 1.00	SUPERPLSTCZR BRAND B1 15.0 -0	.349 .349 228. 3.9 2.0	4.00 152. -0 15. MIN DAMP,73F	(28-DAY) • 8350. 7780. 8240. 9020.	(56-DAY) • 9160. 9660. 8580. 9250.	(28-DAY) • 900. 850. 917. 933.	(28-DAY) • 8490. 8390. 8510. 8590.	(28-DAY) • -0• -0 -0 -0
CBB B 10/12/82 7.0/1.5	BRAND C5 II 663. 12.	 NONE 0 0 0	3/4 LIMESTONE BRAND B2 1938. 44. 75.	2.57 BRAND B2 1303. 30. 1.49	SUPERPLSTCZR BRAND B1 15.0 -0	.296 .296 196. 3.3 2.0	4.00 153. 77.0 15. MIN DAMP,73F	(28-DAY) • 8590. 8540. 8470. 8750.	(56-DAY) • 9000. 9050. 8380. 9570.	(28-DAY) • 980. 1008. 1008. 925.	(28-DAY) • 8670. 8320. 8960. 8720.	(28-DAY) • -0• -0 -0 -0
CBB B 10/11/82 7.0/2.0	BRAND C5 II 661. 13.	 NONE 0 0 0	3/4 LIMESTONE BRAND B2 2131. 48. 82.	2.57 BRAND B2 1058. 24. 2.01	SUPERPLSTCZR BRAND B1 15.1 -0	.331 .331 219. 3.7 2.0	6.25 156. 78.8 15. MIN DAMP,73F	(28-DAY) • 8960. 8890. 8930. 9050.	(56-DAY) • 8870. 9350. 6420. 8380.	(28-DAY) • 993. 892. 1025. 1063.	(28-DAY) • 8660. 8210. 8820. 8940.	(28-DAY) • -0• -0 -0 -0
CBB B 10/19/82 8.5/1.0	BRAND C5 II 807. 15.	 NONE 0 0 0	3/4 LIMESTONE BRAND B2 1498. 34. 58.	2.85 BRAND B3 1500. 35. 1.00	SUPERPLSTCZR BRAND B1 9.6 -0	.300 .300 242. 3.4 2.0	6.00 151. 78.8 15. MIN DAMP,73F	(28-DAY) • 8290. 8450. 8030. 8400.	(56-DAY) • 9460. 9320. 9690. 9370.	(28-DAY) • 986. 967. 1017. 975.	(28-DAY) • 8130. 8180. 7910. 8290.	(28-DAY) • -0• -0 -0 -0
CBB B 10/19/82 8.5/1.5 (M)	BRAND C5 II 809. 15.	 NONE 0 0 0	3/4 LIMESTONE BRAND B2 1804. 41. 70.	2.85 BRAND B3 1203. 28. 1.50	SUPERPLSTCZR BRAND B1 9.7 -0	.297 .297 240. 3.3 2.0	5.25 153. 78.8 15. MIN DAMP,73F	(28-DAY) • 8650. 8790. 8630. 8540.	(56-DAY) • 9590. 9810. 9580. 9390.	(28-DAY) • 978. 1059. 933. 950.	(28-DAY) • 8070. 8820. 6940. 8440.	(28-DAY) • -0• -0 -0 -0

MIX I.D. / MIX DATE / CF/CAFA	CEMENT: BRAND / TYPE / LBS/CUYD / PCT VOLUME	FLYASH: BRAND / CLASS / LBS/CUYD / PCT VOLUME / PCT REPLACED	COARSE AGG: SIZE / MATERIAL / SOURCE / LBS/CUYD / P:VOL,DRUM	FINE AGG: FINENESS / SOURCE / LBS/CUYD / PCT VOLUME / CA/FA(LB/LB)	ADMIXTURE: TYPE / BRAND / DOSE(OZ/100) / (2ND TYPE) / (2ND DOSE)	WATER: W/C / W/B / LB/CUYD / GAL/SACK / PCT AIR	MISC: SLUMP / UNIT WT / MIX TMP / MX TIME / CURING	TEST RESULTS: 6 X 12 CYLINDER (STEEL) (PSI)	6 X 12 CYLINDER (STEEL) (PSI)	6X6X18 BEAM (STEEL) (PSI)	4 X 8 CYLINDER (CARDBD) (PSI)	4 X 8 CYLINDER (STEEL) (PSI)
CFB B	BRAND C5		3/4	2.57	SUPERPLSTCZR	.309	4.25	(28-DAY)	(56-DAY)	(28-DAY)	(28-DAY)	(28-DAY)
10/14/82	II	NONE	LIMESTONE	BRAND B2	BRAND B1	.309	152.	• 8190. •	8590. •	922. •	7960. •	-0•
8.5/2.0	807.	0	BRAND B2	990.	8.1	250.	77.9	8080.	8910.	933.	8420.	-0
	15.	0	1999.	23.		3.5	15. MIN	8350.	8700.	883.	7810.	-0
		0	45. 77.	2.02	-0	2.0	DAMP,73F	8130.	8150.	950.	7640.	-0
CBB B	BRAND C5		3/4	2.85	SUPERPLSTCZR	.307	4.00	(28-DAY)	(56-DAY)	(28-DAY)	(28-DAY)	(28-DAY)
10/21/82	II	NONE	LIMESTONE	BRAND B3	BRAND B1	.307	152.	• 8570. •	8840. •	992. •	8750. •	-0•
10.0/1.0	945.	0	BRAND B2	1379.	6.6	291.	-0	8610.	9940.	1017.	8940.	-0
	18.	0	1378.	32.		3.5	15. MIN	8670.	8930.	983.	8570.	-0
		0	31. 53.	1.00	-0	2.0	DAMP,73F	8440.	8560.	975.	8750.	-0
CBB B	BRAND C5		3/4	2.57	SUPERPLSTCZR	.310	7.25	(28-DAY)	(56-DAY)	(28-DAY)	(28-DAY)	(28-DAY)
10/14/82	II	NONE	LIMESTONE	BRAND B2	BRAND B1	.310	149.	• 7460. •	8130. •	889. •	7190. •	-0•
10.0/1.5	947.	0	BRAND B2	1094.	6.1	294.	77.9	7730.	7980.	925.	7100.	-0
	18.	0	1660.	25.		3.5	15. MIN	7370.	8130.	925.	7420.	-0
		0	37. 64.	1.52	-0	2.0	DAMP,73F	7270.	8290.	817.	7050.	-0
CBB B	BRAND C5		3/4	2.57	SUPERPLSTCZR	.300	4.25	(28-DAY)	(56-DAY)	(28-DAY)	(28-DAY)	(28-DAY)
10/13/82	II	NONE	LIMESTONE	BRAND B2	BRAND B1	.300	153.	• 8220. •	8960. •	990. •	8270. •	-0•
10.0/2.0	952.	0	BRAND B2	914.	6.0	286.	78.8	8290.	8860.	1037.	8070.	-0
	18.	0	1861.	21.		3.4	15. MIN	8120.	9210.	982.	8260.	-0
		0	42. 72.	2.04	-0	2.0	DAMP,73F	8260.	8820.	950.	8480.	-0
CBB B	BRAND C5		3/4	2.57	SUPERPLSTCZR	.339	5.75	(28-DAY)	(56-DAY)	(28-DAY)	(28-DAY)	(28-DAY)
10/18/82	II	NONE	LIMESTONE	BRAND B2	BRAND B1	.339	150.	• 7830. •	8570. •	959. •	7580. •	-0•
8.5/1.5	798.	0	BRAND B2	1160.	9.2	271.	77.9	7670.	8740.	967.	6990.	-0
(M)	15.	0	1777.	27.		3.8	15. MIN	8170.	8440.	942.	7960.	-0
		0	40. 69.	1.53	-0	2.0	DAMP,73F	7830.	8540.	967.	7800.	-0
CFC B	BRAND C5		1/2	2.45	SUPERPLSTCZR	.437	4.00	(28-DAY)	(70-DAY)	(28-DAY)	(28-DAY)	(28-DAY)
10/27/82	II	NONE	GRAVEL	BRAND C2	BRAND B1	.437	151.	• 7290. •	8300. •	708. •	7350. •	8010.•
7.0/1.5	628.	0	BRAND F2	1234.	15.0	274.	-0	7040.	8350.	758.	7660.	8270.
	12.	0	1832.	28.		4.9	15. MIN	7340.	8010.	667.	7440.	8020.
		0	42. 70.	1.48	-0	2.0	DAMP,73F	7500.	8540.	700.	6940.	7730.
CFC B	BRAND C5		1/2	2.45	SUPERPLSTCZR	.281	5.50	(32-DAY)	(70-DAY)	(32-DAY)	(32-DAY)	(32-DAY)
10/28/82	II	NONE	GRAVEL	BRAND C2	BRAND B1	.281	154.	• 9900. •	10240. •	1048. •	9770. •	10890.•
8.5/1.5	807.	0	BRAND F2	1226.	16.7	227.	-0	9660.	10270.	1083.	9180.	10900.
	15.	0	1815.	28.		3.2	15. MIN	10030.	9810.	1042.	9910.	10700.
		0	42. 69.	1.48	-0	2.0	DAMP,73F	10010.	10650.	1020.	10230.	11080.
CFC B	BRAND C5		1/2	2.45	SUPERPLSTCZR	.282	6.25	(32-DAY)	(70-DAY)	(32-DAY)	(32-DAY)	(32-DAY)
10/28/82	II	NONE	GRAVEL	BRAND C2	BRAND B1	.282	152.	• 8530. •	8490. •	920. •	9070. •	9990.•
10.0/1.5	950.	0	BRAND F2	1134.	11.9	268.	-0	8510.	9000.	863.	9230.	9610.
	18.	0	1681.	25.		3.2	15. MIN	8790.	7940.	955.	9120.	10370.
		0	39. 64.	1.48	-0	2.0	DAMP,73F	8280.	8520.	942.	8850.	9990.

	CEMENT	FLYASH	COARSE AGG	FINE AGG	ADMIXTURE	WATER	MISC	••••••••••••••• TEST RESULTS •••••••••••••••				
MIX I.D. MIX DATE CF/CAFA	BRAND TYPE LBS/CUYD PCT VOLUME	BRAND CLASS LBS/CUYD PCT VOLUME PCT REPLACED	SIZE MATERIAL SOURCE LBS/CUYD P:VOL,DRUW	FINENESS SOURCE LBS/CUYD PCT VOLUME CA/FA(LB/LB)	TYPE BRAND DOSE(OZ/100) (2ND TYPE) (2ND DOSE)	W/C W/B LB/CUYD GAL/SACK PCT AIR	SLUMP UNIT WT MIX TMP MX TIME CURING	6 X 12 CYLINDER (STEEL) (PSI)	6 X 12 CYLINDER (STEEL) (PSI)	6X6X18 BEAM (STEEL) (PSI)	4 X 8 CYLINDER (CARDBD) (PSI)	4 X 8 CYLINDER (STEEL) (PSI)
DLC B	BRAND D1		1/2	2.45	SUPERPLSTCZR	.327	4.75	(28-DAY)	(63-DAY)	(28-DAY)	(28-DAY)	(28-DAY)
11/ 1/82	I	NONE	LIMESTONE	BRAND C2	BRAND B1	.327	152.	• 8330.	• 9110.	• 955.	• 7860.	• 9820.•
8.5/1.5	786.	0	BRAND E3	1207.	14.7	257.	-0	8120.	8490.	996.	7450.	9550.
	15.	0	1816.	27.		3.7	15. MIN	8360.	9410.	936.	8500.	9710.
		0	41. 72.	1.50	-0	2.0	DAMP,73F	8520.	9440.	931.	7640.	10190.
DLC B	BRAND D1		1/2	2.45	SUPERPLSTCZR	.341	5.50	(28-DAY)	(63-DAY)	(28-DAY)	(28-DAY)	(28-DAY)
11/ 3/82	I	NONE	LIMESTONE	BRAND C2	BRAND B1	.341	149.	• 8560.	• 9280.	• 976.	• 8230.	• -0•
8.5/1.0	782.	0	BRAND E3	1493.	14.9	267.	-0	8650.	9600.	988.	9000.	-0
	15.	0	1507.	34.		3.8	15. MIN	8540.	9200.	968.	7850.	-0
		0	34. 60.	1.01	-0	2.0	DAMP,73F	8490.	9040.	972.	7830.	-0
CLBAO	BRAND C5	BRAND A1	1/2	2.85		.430	3.00	(28-DAY)	(57-DAY)	(28-DAY)	(28-DAY)	(28-DAY)
11/ 8/82	II	CLASS C	LIMESTONE	BRAND B3	NONE	.344	152.	• 8920.	• 9620.	• 879.	• 8320.	• 9770.•
7.6/2.0	654.	163.	BRAND E3	963.	0	281.	-0	8820.	10030.	864.	8220.	9870.
	13.	4.	1918.	22.		4.8	15. MIN	8740.	9900.	867.	8770.	9380.
		20.	43. 76.	1.99	-0	2.0	DAMP,73F	9210.	9020.	907.	7960.	10070.
CLBAO	BRAND C5	BRAND A1	1/2	2.85		.371	3.00	(28-DAY)	(56-DAY)	(28-DAY)	(28-DAY)	(28-DAY)
11/ 9/82	II	CLASS C	LIMESTONE	BRAND B3	NONE	.297	152.	• 9160.	• 10130.	• 1009.	• 9230.	• 10710.•
8.0/2.0	763.	191.	BRAND E3	927.	0	283.	-0	9320.	10380.	1036.	9110.	10930.
	14.	4.	1832.	21.		4.2	15. MIN	8860.	9670.	1033.	9390.	10350.
		20.	41. 73.	1.98	-0	2.0	DAMP,73F	9300.	10350.	958.	9190.	10840.
CLBAO	BRAND C5	BRAND A1	1/2	2.85		.357	3.00	(28-DAY)	(56-DAY)	(28-DAY)	(28-DAY)	(28-DAY)
11/10/82	II	CLASS C	LIMESTONE	BRAND B3	NONE	.286	152.	• 9450.	• 10620.	• 996.	• 9280.	• 10970.•
8.5/2.0	810.	202.	BRAND E3	897.	0	289.	-0	9740.	10540.	1042.	8990.	11270.
	15.	4.	1776.	21.		4.0	15. MIN	9250.	10700.	1007.	8940.	10900.
		20.	40. 71.	2.00	-0	2.0	DAMP,73F	9370.	10610.	938.	9910.	10740.
CLBAO	BRAND C5	BRAND A1	1/2	2.85		.341	4.00	(28-DAY)	(56-DAY)	(28-DAY)	(28-DAY)	(28-DAY)
11/11/82	II	CLASS C	LIMESTONE	BRAND B3	NONE	.273	151.	• 9360.	• 9950.	• 1011.	• 9640.	• 10890.•
10.0/2.0	958.	239.	BRAND E3	811.	0	327.	-0	9670.	10170.	1028.	9960.	10950.
	18.	5.	1625.	19.		3.8	15. MIN	9320.	9800.	971.	9390.	10970.
		20.	37. 65.	2.00	-0	2.0	DAMP,73F	9090.	9890.	1035.	9580.	10850.
CLBAB	BRAND C5	BRAND A1	1/2	2.85	SUPERPLSTCZR	.344	6.00	(28-DAY)	(64-DAY)	(28-DAY)	(28-DAY)	(28-DAY)
11/15/82	II	CLASS C	LIMESTONE	BRAND B3	BRAND B1	.276	155.	• 9470.	• 10050.	• 1043.	• 10230.	• 11220.•
7.0/2.0	653.	163.	BRAND E3	1010.	15.2	225.	73.0	9670.	9900.	1058.	10380.	11340.
	12.	4.	2021.	23.		3.9	15. MIN	9670.	10420.	1016.	9970.	11300.
		20.	45. 80.	2.00	-0	2.0	DAMP,73F	9070.	9830.	1055.	10350.	11010.
CLBAB	BRAND C5	BRAND A1	1/2	2.85	SUPERPLSTCZR	.321	5.00	(28-DAY)	(63-DAY)	(29-DAY)	(28-DAY)	(28-DAY)
11/16/82	II	CLASS C	LIMESTONE	BRAND B3	BRAND B1	.257	154.	• 9800.	• 10160.	• 1105.	• 9880.	• 10590.•
8.0/2.0	754.	188.	BRAND E3	961.	11.3	242.	73.0	9690.	10350.	1058.	10220.	10900.
	14.	4.	1915.	22.		3.6	15. MIN	9970.	10680.	1173.	9150.	10170.
		20.	43. 76.	1.99	-0	2.0	DAMP,73F	9740.	9460.	1095.	10280.	10700.

	CEMENT	FLYASH	COARSE AGG	FINE AGG	ADMIXTURE	WATER	MISC	********** TEST RESULTS **********				
MIX I.D.	BRAND	BRAND	SIZE	FINENESS	TYPE	W/C	SLUMP					
MIX DATE	TYPE	CLASS	MATERIAL	SOURCE	BRAND	W/B	UNIT WT	6 X 12	6 X 12	6X6X18	4 X 8	4 X 8
CF/CAFA	LBS/CUYD	LBS/CUYD	SOURCE	LBS/CUYD	DOSE(OZ/100)	LB/CUYD	MIX TMP	CYLINDER	CYLINDER	BEAM	CYLINDER	CYLINDER
	PCT VOLUME	PCT VOLUME	LBS/CUYD	PCT VOLUME	(2ND TYPE)	GAL/SACK	MX TIME	(STEEL)	(STEEL)	(STEEL)	(CARDBD)	(STEEL)
		PCT REPLACED	P:VOL,DRUM	CA/FA(LB/LB)	(2ND DOSE)	PCT AIR	CURING	(PSI)	(PSI)	(PSI)	(PSI)	(PSI)
CLBAB	BRAND C5	BRAND A1	1/2	2.85	SUPERPLSTCZR	.316	5.00	(28-DAY)	(62-DAY)	(28-DAY)	(28-DAY)	(28-DAY)
11/19/82	II	CLASS C	LIMESTONE	BRAND B3	BRAND B1	.254	155.	• 9660.	• 10300.	• 1009.	• 9440.	• 10430.•
8.5/2.0	808.	202.	BRAND E4	939.	11.1	256.	75.0	9710.	8790.	1061.	8930.	10730.
(K)	15.	4.	1873.	22.		3.6	15. MIN	9600.	10450.	988.	9410.	10490.
		20.	41. 73.	2.00	-0	2.0	DAMP,73F	8590.	10150.	978.	9980.	10010.
CLBAB	BRAND C5	BRAND A1	1/2	2.85	SUPERPLSTCZR	.325	4.50	(28-DAY)	(63-DAY)	(29-DAY)	(28-DAY)	(28-DAY)
11/18/82	II	CLASS C	LIMESTONE	BRAND B3	BRAND B1	.261	151.	• 9930.	• 10210.	• 1122.	• 9070.	• 10320.•
10.0/2.0	946.	236.	BRAND E4	842.	12.0	308.	75.0	9440.	10290.	1130.	9120.	10250.
	18.	5.	1682.	19.		3.7	15. MIN	10380.	10380.	1075.	8930.	7390.
		20.	37. 66.	2.00	-0	2.0	DAMP,73F	9970.	9970.	1160.	9150.	10380.
CLBAB	BRAND C5	BRAND A1	1/2	2.85	SUPERPLSTCZR	.310	4.75	(28-DAY)	(63-DAY)	(29-DAY)	(28-DAY)	(28-DAY)
11/17/82	II	CLASS C	LIMESTONE	BRAND B3	BRAND B1	.248	155.	• 9890.	• 10550.	• 1119.	• 10600.	• 11340.•
8.5/2.0	810.	202.	BRAND E4	938.	14.4	251.	72.0	10260.	10130.	1120.	10880.	11520.
(K)	15.	4.	1882.	22.		3.5	15. MIN	9620.	10800.	1168.	10190.	11300.
		20.	42. 73.	2.01	-0	2.0	DAMP,73F	9780.	10730.	1070.	10740.	11110.
CLBAO	BRAND C5	BRAND A1	1/2	2.85		.487	3.00	(28-DAY)	(57-DAY)	(28-DAY)	(28-DAY)	(28-DAY)
11/22/82	II	CLASS C	LIMESTONE	BRAND B3	NONE	.341	152.	• 9300.	• 10060.	• 868.	• -0	• 9500.•
5.9/2.0	562.	241.	BRAND E4	976.	0	274.	75.0	9200.	10240.	863.	-0	9610.
	11.	5.	1956.	23.		5.5	15. MIN	9210.	10220.	897.	-0	9770.
		30.	43. 76.	2.00	-0	2.0	DAMP,73F	9500.	9730.	845.	-0	9110.
CLBAO	BRAND C5	BRAND A1	1/2	2.95		.441	3.00	(42-DAY)	(56-DAY)	(42-DAY)	(42-DAY)	(42-DAY)
11/23/82	II	CLASS C	LIMESTONE	BRAND B3	NONE	.308	150.	• 9950.	• 10070.	• 957.	• -0	• 10930.•
7.0/2.0	657.	281.	BRAND E4	919.	0	289.	76.0	9670.	9990.	987.	-0	11190.
(S)	12.	6.	1853.	21.		5.0	15. MIN	10360.	10490.	903.	-0	11340.
		30.	41. 72.	2.02	-0	2.0	DAMP,73F	9830.	9730.	980.	-0	9950.
CLBAO	BRAND C5	BRAND A1	1/2	2.85		.404	3.00	(35-DAY)	(58-DAY)	(35-DAY)	(35-DAY)	(35-DAY)
11/30/82	II	CLASS C	LIMESTONE	BRAND B3	NONE	.283	151.	• 9450.	• 9480.	• 1103.	• -0	• 10730.•
8.5/2.0	803.	344.	BRAND E4	826.	0	325.	76.0	9640.	9370.	1124.	-0	10150.
	15.	8.	1668.	19.		4.6	15. MIN	9500.	9780.	1056.	-0	11460.
		30.	37. 65.	2.02	-0	2.0	DAMP,73F	9210.	9280.	1129.	-0	10570.
CLBAO	BRAND C5	BRAND A1	1/2	2.85		.368	3.00	(35-DAY)	(58-DAY)	(35-DAY)	(35-DAY)	(35-DAY)
11/30/82	II	CLASS C	LIMESTONE	BRAND B3	NONE	.257	150.	• 9430.	• 9400.	• 1154.	• -0	• 10860.•
10.0/2.0	949.	407.	BRAND E4	745.	0	349.	75.0	8980.	9460.	1118.	-0	10420.
	18.	9.	1499.	17.		4.1	15. MIN	9410.	9800.	1161.	-0	10890.
		30.	33. 58.	2.01	-0	2.0	DAMP,73F	9900.	8950.	1183.	-0	11280.
CLBAB	BRAND C5	BRAND A1	1/2	2.85	SUPERPLSTCZR	.388	5.50	(35-DAY)	(57-DAY)	(35-DAY)	(35-DAY)	(35-DAY)
12/ 1/82	II	CLASS C	LIMESTONE	BRAND B3	BRAND B1	.272	155.	•11910.	• 11030.	• 988.	• -0	• 11330.•
5.9/2.0	556.	239.	BRAND E4	1033.	17.5	216.	75.0	11390.	11260.	963.	-0	10790.
(T)	10.	5.	2058.	24.		4.4	15. MIN	11070.	11000.	969.	-0	11520.
		30.	46. 80.	1.99	-0	2.0	DAMP,73F	10560.	10840.	1033.	-0	11670.

	CEMENT	FLYASH	COARSE AGG	FINE AGG	ADMIXTURE	WATER	MISC	************** TEST RESULTS **************				
MIX I.D. MIX DATE CF/CAFA	BRAND TYPE LBS/CUYD PCT VOLUME	BRAND CLASS LBS/CUYD PCT VOLUME PCT REPLACED	SIZE MATERIAL SOURCE LBS/CUYD P:VOL,DRUW	FINENESS SOURCE LBS/CUYD PCT VOLUME CA/FA(LB/LB)	TYPE BRAND DOSE(OZ/100) (2ND TYPE) (2ND DOSE)	W/C W/B LB/CUYD GAL/SACK PCT AIR	SLUMP UNIT WT MIX TMP MX TIME CURING	6 X 12 CYLINDER (STEEL) (PSI)	6 X 12 CYLINDER (STEEL) (PSI)	6X6X18 BEAM (STEEL) (PSI)	4 X 8 CYLINDER (CARDBD) (PSI)	4 X 8 CYLINDER (STEEL) (PSI)
C:BAB 12/ 2/82 7.0/2.0	BRAND C5 II 660. 12.	BRAND A1 CLASS C 282. 6. 30.	1/2 LIMESTONE BRAND E4 1929. 43. 75.	2.85 BRAND B3 963. 22. 2.00	SUPERPLSTCZR BRAND B1 12.6 -0	.367 .257 242. 4.1 2.0	5.25 154. 74.0 15. MIN DAMP,73F	(35-DAY) *10090. 10400. 9890. 9990.	(56-DAY) * 10030. 9990. 10200. 9890.	(35-DAY) * 1097. 1170. 1071. 1051.	(35-DAY) * -0 -0 -0 -0	(35-DAY) * 10730.* 10330. 10920. 10930.
C:BAB 12/ 2/82 8.5/2.0	BRAND C5 II 803. 15.	BRAND A1 CLASS C 344. 8. 30.	1/2 LIMESTONE BRAND E4 1722. 38. 67.	2.85 BRAND B3 862. 20. 2.00	SUPERPLSTCZR BRAND B1 10.9 -0	.361 .253 290. 4.1 2.0	5.00 151. 76.0 15. MIN DAMP,73F	(35-DAY) * 9560. 9870. 9710. 9370.	(56-DAY) * 10560. 10330. 10790. 9500.	(35-DAY) * 1130. 1197. 1083. 1111.	(35-DAY) * -0 -0 -0 -0	(35-DAY) * 10420.* 11110. 10070. 10090.
C'BAB 12/ 3/82 10.0/2.0	BRAND C5 II 947. 18.	BRAND A1 CLASS C 405. 9. 30.	1/2 LIMESTONE BRAND E4 1515. 34. 59.	2.85 BRAND B3 761. 18. 1.99	SUPERPLSTCZR BRAND B1 6.0 -0	.357 .250 338. 4.0 2.0	4.00 150. 74.0 15. MIN DAMP,73F	(35-DAY) * 9120. 9570. 9040. 8740.	(60-DAY) * 9660. 9500. 9440. 10030.	(35-DAY) * 1078. 1061. 1093. 1081.	(35-DAY) * -0 -0 -0 -0	(35-DAY) * 10510.* 10170. 10590. 10680.
C-BAO 12/ 6/82 10.0/2.5	BRAND C5 II 924. 17.	BRAND A1 CLASS C 231. 5. 20.	1/2 LIMESTONE BRAND E4 1783. 39. 78.	2.85 BRAND B3 710. 16. 2.51	NONE 0 -0	.358 .286 330. 4.0 2.0	3.25 151. 74.0 15. MIN DAMP,73F	(28-DAY) * 9360. 9350. 9730. 9000.	(57-DAY) * 9560. 9500. 9890. 9600.	(28-DAY) * 1048. 1105. 1032. 1008.	(28-DAY) * -0 -0 -0 -0	(28-DAY) * 10300.* 9990. 10120. 10780.
CIBAB 12/ 7/82 10.0/2.5	BRAND C5 II 941. 18.	BRAND A1 CLASS C 235. 5. 20.	1/2 LIMESTONE BRAND E4 1833. 41. 71.	2.85 BRAND B3 735. 17. 2.49	SUPERPLSTCZR BRAND B1 9.6 -0	.313 .251 295. 3.5 2.0	4.00 154. 74.0 15. MIN DAMP,73F	(28-DAY) * 9330. 9550. 8770. 9660.	(56-DAY) * 9540. 9710. 9300. 9600.	(28-DAY) * 1095. 1102. 1167. 1017.	(28-DAY) * -0 -0 -0 -0	(28-DAY) * 10660.* 11190. 9970. 11090.
CIBAO 12/15/82 10.0/3.0	BRAND C5 II 919. 17.	BRAND A2 CLASS C 230. 5. 20.	1/2 LIMESTONE BRAND E4 1869. 41. 73.	2.85 BRAND B3 615. 14. 3.04	NONE 0 -0	.365 .292 335. 4.1 2.0	3.00 152. 71.0 15. MIN DAMP,73F	(35-DAY) * 9160. 9070. 9600. 8920.	(59-DAY) * 9940. 10220. 9830. 9780.	(35-DAY) * 898. 878. 933. 883.	(35-DAY) * -0 -0 -0 -0	(35-DAY) * 9340.* 8850. 9680. 9490.
CIBAB 12/15/82 10.0/3.0	BRAND C5 II 938. 18.	BRAND A2 CLASS C 235. 5. 20.	1/2 LIMESTONE BRAND E4 1930. 43. 75.	2.85 BRAND B3 635. 15. 3.04	SUPERPLSTCZR BRAND B1 9.7 -0	.317 .253 297. 3.6 2.0	4.50 154. 71.5 15. MIN DAMP,73F	(35-DAY) * 8970. 8840. 9020. 9050.	(59-DAY) * 9060. 8930. 9050. 9290.	(35-DAY) * 1104. 1090. 1174. 1047.	(35-DAY) * -0 -0 -0 -0	(35-DAY) * 10730.* 10570. 10900. 10710.
C'BAO 12/17/82 10.0/2.5	BRAND C5 II 956. 18.	BRAND A2 CLASS C 410. 9. 30.	1/2 LIMESTONE BRAND E4 1582. 35. 62.	2.85 BRAND B3 631. 15. 2.51	NONE 0 -0	.372 .261 356. 4.2 2.0	3.00 147. 71.0 15. MIN DAMP,73F	(34-DAY) * 9300. 9300. 9180. 9410.	(57-DAY) * 10240. 9900. 10430. 10380.	(34-DAY) * 1084. 1065. 1029. 1155.	(34-DAY) * -0 -0 -0 -0	(34-DAY) * 10400.* 9680. 10630. 10890.

MIX I.D. MIX DATE CF/CAFA	CEMENT BRAND TYPE LBS/CUYD PCT VOLUME	FLYASH BRAND CLASS LBS/CUYD PCT VOLUME PCT REPLACED	COARSE AGG SIZE MATERIAL SOURCE LBS/CUYD P:VOL,DRUW	FINE AGG FINENESS SOURCE LBS/CUYD PCT VOLUME CA/FA(LB/LB)	ADMIXTURE TYPE BRAND DOSE(OZ/100) (2ND TYPE) (2ND DOSE)	WATER W/C W/B LB/CUYD GAL/SACK PCT AIR	MISC SLUMP UNIT WT MIX TMP MX TIME CURING	TEST RESULTS 6 X 12 CYLINDER (STEEL) (PSI)	6 X 12 CYLINDER (STEEL) (PSI)	6X6X18 BEAM (STEEL) (PSI)	4 X 8 CYLINDER (CARDBD) (PSI)	4 X 8 CYLINDER (STEEL) (PSI)
CFBA0	BRAND C6	BRAND A2	1/2	2.85		.378	3.00	(28-DAY)	(56-DAY)	(28-DAY)	(28-DAY)	(28-DAY)
1/18/83	II	CLASS C	GRAVEL	BRAND B3	NONE	.302	150.	• 8820.	∘ 9730.	∘ 1032.	• -0	• 10150.•
8.5/2.0	790.	197.	BRAND F2	876.	0	298.	71.0	8790.	9410.	991.	-0	10170.
	15.	4.	1769.	20.		4.3	15. MIN	8670.	9940.	1074.	-0	10090.
		20.	41. 67.	2.02	-0	2.0	DAMP,73F	9000.	9830.	1032.	-0	10250.
CFBAB	BRAND C6	BRAND A2	1/2	2.85	SUPERPLSTCZR	.322	4.00	(28-DAY)	(56-DAY)	(28-DAY)	(28-DAY)	(28-DAY)
1/20/83	II	CLASS C	GRAVEL	BRAND B3	BRAND B1	.257	153.	• 9040.	• 9730.	• 1100.	• -C	• 9870.•
8.5/2.0	794.	198.	BRAND F2	923.	9.6	255.	70.0	9000.	9830.	1073.	-0	9510.
	15.	4.	1829.	21.		3.6	15. MIN	9480.	9460.	1110.	-0	10110.
		20.	42. 70.	1.98	-0	2.0	DAMP,73F	8650.	9890.	1117.	-0	9980.
CHBA0	BRAND C6	BRAND A2	3/4	2.85		.375	3.75	(28-DAY)	(56-DAY)	(28-DAY)	(28-DAY)	(28-DAY)
1/19/83	II	CLASS C	LIMESTONE	BRAND B3	NONE	.300	151.	• 9460.	∘ 9450.	• 1044.	• -0	∘ 10470.•
8.5/2.0	784.	196.	BRAND B2	895.	0	294.	70.0	9670.	9270.	1108.	-0	10540.
	15.	4.	1801.	21.		4.2	15. MIN	9530.	9640.	1073.	-0	10450.
		20.	41. 69.	2.01	-0	2.0	DAMP,73F	9180.	9430.	952.	-0	10420.
CHBAB	BRAND C6	BRAND A2	3/4	2.85	SUPERPLSTCZR	.327	7.50	(28-DAY)	(56-DAY)	(28-DAY)	(28-DAY)	(28-DAY)
1/21/83	II	CLASS C	LIMESTONE	BRAND B3	BRAND B1	.261	154.	• 9600.	• 9770.	∘ 1169.	• -0	• 10900.•
8.5/2.0	788.	196.	BRAND B2	924.	11.2	257.	71.0	9510.	9710.	1201.	-0	10630.
	15.	4.	1865.	21.		3.7	15. MIN	9600.	9970.	1088.	-0	10970.
		20.	42. 72.	2.02	-0	2.0	DAMP,73F	9690.	9730.	1217.	-0	11160.
C.BAB	BRAND C6	BRAND A2	1/2	2.85	SUPERPLSTCZR	.424	7.50	(28-DAY)	(56-DAY)	(28-DAY)	(28-DAY)	(28-DAY)
1/26/83	II	CLASS C	LIMESTONE	BRAND B3	BRAND B1	.298	152.	•10170.	∘ 10940.	• 917.	• -0	• 10970.•
7../2.0	652.	273.	BRAND E4	934.	13.1	277.	73.0	9020.	10630.	888.	-0	10950.
	12.	6.	1869.	22.		4.8	15. MIN	10170.	11000.	975.	-0	11400.
		30.	41. 73.	2.00	-0	2.0	DAMP,73F	8520.	11180.	888.	-0	10570.
C:B B	BRAND C6		1/2	2.85	SUPERPLSTCZR	.312	8.00	(28-DAY)	(56-DAY)	(28-DAY)	(28-DAY)	(28-DAY)
1/26/83	II	NONE	LIMESTONE	BRAND B3	BRAND B1	.312	155.	• 9000.	• 9170.	∘ 1015.	• -0	• 10380.•
10.0/2.0	934.	0	BRAND E4	936.	10.4	291.	72.0	8890.	9230.	996.	-0	10270.
	18.	0	1874.	22.		3.5	15. MIN	8560.	9320.	1038.	-0	10100.
		0	42. 73.	2.00	-0	2.0	DAMP,73F	9550.	8970.	1012.	-0	10770.
CBR 0	BRAND C6		3/4	2.85	SUPERPLSTCZR	.321	4.00	(28-DAY)	(55-DAY)	(28-DAY)	(28-DAY)	(28-DAY)
1/28/83	II	NONE	LIMESTONE	BRAND B3	BRAND B2	.321	152.	• 8820.	• 9210.	• 895.	• -0	• 9820.•
8.5/1.5	791.	0	BRAND B2	1180.	9.7	254.	73.0	8670.	9180.	920.	-0	9640.
(3/4" Fine)	15.	0	1807.	27.		3.6	15. MIN	8680.	9530.	878.	-0	9960.
		0	41. 70.	1.53	-0	2.0	DAMP,73F	9110.	8910.	888.	-0	9850.
CHR B	BRAND C6		3/4	2.85	SUPERPLSTCZR	.309	4.25	(28-DAY)	(55-DAY)	(28-DAY)	(28-DAY)	(28-DAY)
1/28/83	II	NONE	LIMESTONE	BRAND B3	BRAND B2	.309	154.	• 8630.	• 9470.	• 956.	• -0	• 9890.•
8.5/1.5	795.	0	BRAND B2	1187.	9.7	246.	73.0	8400.	8360.	1000.	-0	10030.
(3/4" Medium)	15.	0	1817.	27.		3.5	15. MIN	8970.	9390.	907.	-0	9850.
		0	41. 70.	1.53	-0	2.0	DAMP,73F	8520.	9550.	960.	-0	9790.

MIX I.D. MIX DATE CF/CAFA	CEMENT BRAND TYPE LBS/CUYD PCT VOLUME	FLYASH BRAND CLASS LBS/CUYD PCT VOLUME PCT REPLACED	COARSE AGG SIZE MATERIAL SOURCE LBS/CUYD P:VOL,DRUW	FINE AGG FINENESS SOURCE LBS/CUYD PCT VOLUME CA/FA(LB/LB)	ADMIXTURE TYPE BRAND DOSE(OZ/100) (2ND TYPE) (2ND DOSE)	WATER W/C W/B LB/CUYD GAL/SACK PCT AIR	MISC SLUMP UNIT WT MIX TMP MX TIME CURING	TEST RESULTS 6 X 12 CYLINDER (STEEL) (PSI)	 6 X 12 CYLINDER (STEEL) (PSI)	 6X6X18 BEAM (STEEL) (PSI)	 4 X 8 CYLINDER (CARDBD) (PSI)	 4 X 8 CYLINDER (STEEL) (PSI)
CBB 8	BRAND C6		3/4	2.85	SUPERPLSTCZR	.305	5.50	(28-DAY)	(60-DAY)	(29-DAY)	(28-DAY)	(28-DAY)
1/29/83	II	NONE	LIMESTONE	BRAND B3	BRAND B2	.305	153.	* 9060. *	9810. *	900. *	-0 *	9860.*
8.5/1.5	797.	0	BRAND B2	1190.	9.7	243.	73.0	8790.	9780.	903.	-0	9710.
(3/4" Coarse)	15.	0	1820.	27.		3.4	15. MIN	9200.	9900.	843.	-0	98[illegible]
		0	41. 70.	1.53	-0	2.0	DAMP,73F	9180.	9760.	955.	-0	10010.
E.C 0	BRAND E2		1/2	2.45	SUPERPLSTCZR	.371	5.75	(28-DAY)	(60-DAY)	(28-DAY)	(28-DAY)	(28-DAY)
1/29/83	III	NONE	LIMESTONE	BRAND C2	BRAND B2	.371	150.	* 9470. *	9250. *	996. *	-0 *	10200.*
8.5/1.0	791.	0	BRAND E4	1466.	15.0	290.	75.0	9000.	8880.	1044.	-0	10090.
	15.	0	1496.	33.		4.2	15. MIN	9620.	9600.	903.	-0	10290.
		0	33. 58.	1.02	-0	2.0	DAMP,73F	9780.	9270.	1042.	-0	10270.
F:C 8	BRAND E2		1/2	2.45	SUPERPLSTCZR	.375	4.00	(28-DAY)	(58-DAY)	(29-DAY)	(28-DAY)	(28-DAY)
1/31/83	III	NONE	LIMESTONE	BRAND C2	BRAND B2	.375	153.	* 8930. *	9860. *	1004. *	-0 *	10310.*
8.5/1.5	779.	0	BRAND E4	1165.	15.0	292.	76.0	8540.	9960.	1003.	-0	10030.
	15.	0	1796.	26.		4.2	15. MIN	8890.	10190.	1052.	-0	10420.
		0	40. 70.	1.54	-0	2.0	DAMP,73F	9350.	9430.	958.	-0	10490.
C:BBB	BRAND C6	BRAND B1	1/2	2.85	SUPERPLSTCZR	.390	5.00	(28-DAY)	(58-DAY)	(28-DAY)	(28-DAY)	(28-DAY)
1/31/83	II	CLASS C	LIMESTONE	BRAND B3	BRAND B2	.266	155.	*10290. *	11320. *	921. *	-0 *	11370.*
5.5/2.0	548.	235.	BRAND E4	1033.	18.1	208.	72.5	9940.	11830.	957.	-0	116[illegible]
	10.	6.	2074.	24.		4.3	15. MIN	10790.	11210.	860.	-0	11480.
		30.	46. 81.	2.01	-0	2.0	DAMP,73F	10130.	10930.	947.	-0	11000.
C:BB0	BRAND C6	BRAND B1	1/2	2.85		.393	3.25	(28-DAY)	(56-DAY)	(29-DAY)	(28-DAY)	(28-DAY)
2/ 2/83	II	CLASS C	LIMESTONE	BRAND B3	NONE	.314	150.	* 9520. *	9840. *	852. *	-0 *	10420.*
8.5/2.0	789.	138.	BRAND E4	894.	0	310.	71.5	9800.	9640.	817.	-0	10190.
	15.	5.	1781.	21.		4.4	15. MIN	9200.	9760.	885.	-0	10460.
		20.	39. 69.	1.99	-0	2.0	DAMP,73F	9570.	10120.	855.	-0	10620.
C:BB0	BRAND C6	BRAND B1	1/2	2.85		.466	3.88	(28-DAY)	(56-DAY)	(28-DAY)	(28-DAY)	(28-DAY)
2/ 2/83	II	CLASS C	LIMESTONE	BRAND B3	NONE	.326	151.	* 9030. *	10070. *	798. *	-0 *	10190.*
7.0/2.0	653.	280.	BRAND E4	908.	0	304.	71.0	9320.	10010.	813.	-0	10580.
	12.	7.	1809.	21.		5.2	15. MIN	8750.	10380.	757.	-0	9710.
		30.	40. 71.	1.93	-0	2.0	DAMP,73F	9020.	9810.	825.	-0	10280.
C:B 0	BRAND C6		1/2	2.85		.366	3.00	(28-DAY)	(59-DAY)	(28-DAY)	(28-DAY)	(28-DAY)
2/11/83	II	NONE	LIMESTONE	BRAND B3	NONE	.365	152.	* 8910. *	9460. *	-0 *	-0 *	-0*
10.0/2.1	921.	0	BRAND E4	866.	0	337.	69.0	8910.	9430.	-0	-0	-0
(Q)	17.	0	1834.	20.		4.1	15. MIN	-0	9830.	-0	-0	-0
		0	41. 72.	2.12	-0	2.0	DAMP,73F	-0	9120.	-0	-0	-0

ADDITIONAL DATA FROM MIX ABOVE:

1W	6X12 COMPR STEEL 1D	* 4270.*	4260.	4090.	4460.
2W	6X12 COMPR STEEL 7D	* 7440.*	7220.	7430.	7660.
3W	6X12 SPLIT STEEL 28D	* 831.*	867.	848.	778.

	CEMENT	FLYASH	COARSE AGG	FINE AGG	ADMIXTURE	WATER	MISC	••••••••••••••• TEST RESULTS	••••••••••••••			
MIX I.D. MIX DATE CF/CAFA	BRAND TYPE LBS/CUYD PCT VOLUME	BRAND CLASS LBS/CUYD PCT VOLUME PCT REPLACED	SIZE MATERIAL SOURCE LBS/CUYD P:VOL,DRUM	FINENESS SOURCE LBS/CUYD PCT VOLUME CA/FA(LB/LB)	TYPE BRAND DOSE(OZ/100) (2ND TYPE) (2ND DOSE)	W/C W/B LB/CUYD GAL/SACK PCT AIR	SLUMP UNIT WT MIX TMP MX TIME CURING	6 X 12 CYLINDER (STEEL) (PSI)	6 X 12 CYLINDER (STEEL) (PSI)	6X6X18 BEAM (STEEL) (PSI)	4 X 8 CYLINDER (CARDBD) (PSI)	4 X 8 CYLINDER (STEEL) (PSI)
C:B B	BRAND C6		1/2	2.85	SUPERPLSTCZR	.315	3.50	(28-DAY)	(58-DAY)	(28-DAY)	(28-DAY)	(28-DAY
2/12/83	II	NONE	LIMESTONE	BRAND B3	BRAND B2	.315	156.	•10610. •	11130. •	-0 •	-0 •	-0•
8.5/2.0	785.	0	BRAND E4	1011.	14.9	247.	71.0	10420.	11420.	-0	-0	-0
	15.	0	2041.	23.		3.5	15. MIN	11110.	11420.	-0	-0	-0
(R)		0	45. 80.	2.02	-0	2.0	DAMP,73F	10290.	10560.	-0	-0	-0
ADDITIONAL DATA FROM MIX ABOVE:			1# 6X12	COMPR STEEL 1D	• 6300.•	6350.	6120.	6440.				
			2# 6X12	COMPR STEEL 7D	• 8860.•	8820.	8650.	9120.				
			3# 6X12	SPLIT STEEL 28D	• 801.•	675.	817.	912.				
C:BAO	BRAND C6	BRAND A2	1/2	2.85		.462	2.75	(28-DAY)	(56-DAY)	(28-DAY)	(28-DAY)	(28-DAY)
2/14/83	II	CLASS C	LIMESTONE	BRAND B3	NONE	.323	152.	• 9630. •	10190. •	-0 •	-0 •	-0•
7.0/2.0	653.	280.	BRAND E4	916.	0	301.	70.5	9280.	10220.	-0	-0	-0
	12.	6.	1821.	21.		5.2	15. MIN	9640.	10590.	-0	-0	-0
(S)		30.	40. 71.	1.99	-0	2.0	DAMP,73F	9970.	9740.	-0	-0	-0
ADDITIONAL DATA FROM MIX ABOVE:			1# 6X12	COMPR STEEL 1D	• 4120.•	3980.	4070.	4300.				
			2# 6X12	COMPR STEEL 7D	• 7350.•	7960.	7960.	7920.				
			3# 6X12	SPLIT STEEL 28D	• 799.•	889.	746.	762.				
C:BAB	BRAND C6	BRAND A2	1/2	2.85	SUPERPLSTCZR	.396	4.00	(28-DAY)	(57-DAY)	(28-DAY)	(28-DAY)	(28-DAY)
2/15/83	II	CLASS C	LIMESTONE	BRAND B3	BRAND B2	.277	156.	•11640. •	11340. •	-0 •	-0 •	-0•
5.4/2.0	553.	237.	BRAND E4	1040.	23.0	219.	72.5	11690.	12010.	-0	-0	-0
	10.	5.	2042.	24.		4.5	15. MIN	11620.	10840.	-0	-0	-0
(T)		30.	45. 80.	1.96	-0	2.0	DAMP,73F	11600.	11180.	-0	-0	-0
ADDITIONAL DATA FROM MIX ABOVE:			1# 6X12	COMPR STEEL 1D	• 5900.•	4850.	5910.	5890.				
			2# 6X12	COMPR STEEL 7D	• 9520.•	9530.	9620.	9410.				
			3# 6X12	SPLIT STEEL 28D	• 844.•	761.	993.	778.				
C:B O	BRAND C6		1/2	2.85		.349	2.75	(28-DAY)	(-0-DAY)	(28-DAY)	(28-DAY)	(28-DAY)
2/16/83	II	NONE	LIMESTONE	BRAND B3	NONE	.349	150.	• 8890. •	-0 •	-3 •	8640. •	9810.•
10.0/2.0	930.	0	BRAND E4	901.	0	324.	72.0	8510.	-0	-0	8190.	9760.
	18.	0	1825.	21.		3.9	15. MIN	8930.	-0	-0	8720.	9800.
(Q)		0	40. 71.	2.03	-0	2.0	DAMP,73F	8840.	-0	-0	9020.	9880.
ADDITIONAL DATA FROM MIX ABOVE:			1# 6X12	COMPR PLSTC 28D	• 8230.•	7960.	8280.	8440.				
			2# 6X12	COMPR CRDBD 28D	• 8490.•	8440.	8510.	8510.				
			3# 6X12	SPLIT CRDBD 28D	• 738.•	754.	701.	760.				

MIX I.D. MIX DATE CF/CAFA	CEMENT BRAND TYPE LBS/CUYD PCT VOLUME	FLYASH BRAND CLASS LBS/CUYD PCT VOLUME PCT REPLACED	COARSE AGG SIZE MATERIAL SOURCE LBS/CUYD P:VOL,DRUW	FINE AGG FINENESS SOURCE LBS/CUYD PCT VOLUME CA/FA(LB/LB)	ADMIXTURE TYPE BRAND DOSE(OZ/100) (2ND TYPE) (2ND DOSE)	WATER W/C W/B LB/CUYD GAL/SACK PCT AIR	MISC SLUMP UNIT WT MIX TMP MX TIME CURING	TEST RESULTS 6 X 12 CYLINDER (STEEL) (PSI)	6 X 12 CYLINDER (STEEL) (PSI)	6X6X18 BEAM (STEEL) (PSI)	4 X 8 CYLINDER (CARDBD) (PSI)	4 X 8 CYLINDER (STEEL (PSI)
C:B B 2/18/83 8.5/2.0 (R)	BRAND C6 II 790. 15.	NONE 0 0 0	1/2 LIMESTONE BRAND E4 2046. 45. 80.	2.77 BRAND B4 1043. 24. 1.96	SUPERPLSTCZR BRAND B2 16.7 -0	.290 .290 229. 3.3 2.0	3.75 157. 71.0 15. MIN DAMP,73F	(28-DAY) 9500. 9240. 9320. 9900.	(-0-DAY) -0 -0 -0 -0	(28-DAY) -0 -0 -0 -0	(28-DAY) 10540. 10220. 10900. 10490.	(28-DAY) 11150. 11140. 10580. 11730.
ADDITIONAL DATA FROM MIX ABOVE:			1M 6X12 2M 6X12 3M 6X12	COMPR PLSTC 28D COMPR CRDBD 28D SPLIT CRDBD 28D	10730. 9730. 818.	10500. 9230. 788.	11020. 9900. 822.	10660. 10060. 843.				
C:BAD 2/17/83 7.0/2.0 (S)	BRAND C6 II 662. 12.	BRAND A2 CLASS C 283. 6. 30.	1/2 LIMESTONE BRAND E4 1844. 41. 72.	2.85 BRAND B3 919. 21. 2.01	NONE 0 -0	.435 .305 288. 4.9 2.0	3.00 151. 71.5 15. MIN DAMP,73F	(28-DAY) 9560. 9140. 9480. 10060.	(-0-DAY) -0 -0 -0 -0	(28-DAY) -0 -0 -0 -0	(28-DAY) 10240. 10280. 9840. 10600.	(28-DAY) 10480. 10340. 11060. 10030.
ADDITIONAL DATA FROM MIX ABOVE:			1M 6X12 2M 6X12 3M 6X12	COMPR PLSTC 28D COMPR CRDBD 28D SPLIT CRDBD 28D	8930. 9090. 757.	9070. 9070. 707.	8950. 6700. 749.	8770. 9110. 815.				
C:BAB 2/21/83 5.9/2.0 (T)	BRAND C6 II 556. 10.	BRAND A2 CLASS C 239. 5. 30.	1/2 LIMESTONE BRAND E4 2039. 45. 79.	2.77 BRAND B4 1009. 23. 2.02	SUPERPLSTCZR BRAND B2 18.0 -0	.411 .288 229. 4.6 2.0	7.50 156. 70.0 15. MIN DAMP,73F	(28-DAY) 10210. 9830. 9890. 10910.	(-0-DAY) -0 -0 -0 -0	(28-DAY) -0 -0 -0 -0	(28-DAY) 10330. 10190. 10600. 10190.	(28-DAY) 11080. 10420. 11120. 11710.
ADDITIONAL DATA FROM MIX ABOVE:			1M 6X12 2M 6X12 3M 6X12	COMPR PLSTC 28D COMPR CRDBD 28D SPLIT CRDBD 25D	10960. 10060. 687.	11260. 10360. 592.	10490. 10260. 699.	11140. 9550. 769.				
C:B D 2/25/83 10.0/2.0 (Q)	BRAND C6 II 943. 18.	NONE 0 0 0	1/2 LIMESTONE BRAND E4 1804. 40. 70.	2.77 BRAND B4 902. 21. 2.00	NONE 0 0	.346 .346 326. 3.9 2.0	2.75 -0 106.0 60. MIN DAMP,73F	(27-DAY) 7930. 7600. 7940. 8240.	(-0-DAY) -0 -0 -0 -0	(27-DAY) -0 -0 -0 -0	(27-DAY) -0 -0 -0 -0	(27-DAY) -0 -0 -0 -0
C:B R 3/10/83 8.5/2.0 (R)	BRAND C6 II 798. 15.	NONE 0 0 0	1/2 LIMESTONE BRAND E4 2065. 46. 81.	2.77 BRAND B4 1034. 24. 2.00	SUPERPLSTCZR BRAND B2 15.2 -0	.280 .280 223. 3.1 2.0	4.00 -0 104.0 60. MIN DAMP,73F	(28-DAY) 10400. 9620. 10420. 11180.	(-0-DAY) -0 -0 -0 -0	(28-DAY) -0 -0 -0 -0	(28-DAY) -0 -0 -0 -0	(28-DAY) -0 -0 -0 -0

MIX I.D. MIX DATE CF/CAFA	CEMENT BRAND TYPE LBS/CUYD PCT VOLUME	FLYASH BRAND CLASS LBS/CUYD PCT VOLUME PCT REPLACED	COARSE AGG SIZE MATERIAL SOURCE LBS/CUYD P:VOL,DRUW	FINE AGG FINENESS SOURCE LBS/CUYD PCT VOLUME CA/FA(LB/LB)	ADMIXTURE TYPE BRAND DOSE(OZ/100) (2ND TYPE) (2ND DOSE)	WATER W/C W/B LB/CUYD GAL/SACK PCT AIR	MISC SLUMP UNIT WT MIX TMP MX TIME CURING	TEST RESULTS 6 X 12 CYLINDER (STEEL) (PSI)	6 X 12 CYLINDER (STEEL) (PSI)	6X6X18 BEAM (STEEL) (PSI)	4 X 8 CYLINDER (CARDBD) (PSI)	4 X 8 CYLINDER (STEEL) (PSI)
CIB B 3/10/83 8.5/2.0 (R)	BRAND C6 II 797. 15.	NONE 0 0 0	1/2 LIMESTONE BRAND E4 2064. 46. 80.	2.77 BRAND B4 1033. 24. 2.00	SUPERPLSTCZR BRAND B2 18.0 -0	.282 .282 225. 3.2 2.0	4.50 -0 97.0 90. MIN DAMP,73F	(28-DAY) •11470. • 11440. 11350. 11620.	(-0-DAY) -0 • -0 -0 -0	(28-DAY) -0 • -0 -0 -0	(28-DAY) -0 • -0 -0 -0	(28-DAY -0. -0 -0 -0
C BAO 3/ 2/83 7.0/2.0 (S)	BRAND C6 II 646. 12.	BRAND A2 CLASS C 276. 6. 30.	1/2 LIMESTONE BRAND E4 1803. 40. 70.	2.77 BRAND B4 901. 21. 2.00	NONE 0 -0	.489 .343 316. 5.5 2.0	3.00 -0 108.0 60. MIN DAMP,73F	(28-DAY) • 8490. • 8650. 8470. 8360.	(-0-DAY) -0 • -0 -0 -0	(28-DAY) -0 • -0 -0 -0	(28-DAY) -0 • -0 -0 -0	(28-DAY) -0. -0 -0 -0
CLBAO 3/ 2/83 7.0/2.0 (S)	BRAND C6 II 638. 12.	BRAND A2 CLASS C 273. 6. 30.	1/2 LIMESTONE BRAND E4 1782. 39. 69.	2.77 BRAND B4 891. 21. 2.00	NONE 0 -0	.519 .363 331. 5.8 2.0	3.00 -0 108.0 90. MIN DAMP,73F	(28-DAY) • 8080. • 7590. 7980. 8680.	(-0-DAY) -0 • -0 -0 -0	(28-DAY) -0 • -0 -0 -0	(28-DAY) -0 • -0 -0 -0	(28-DAY) -0. -0 -0 -0
CIBAB 3/ 9/83 5.9/2.0 (T)	BRAND C6 II 565. 11.	BRAND A2 CLASS C 243. 5. 30.	1/2 LIMESTONE BRAND E4 2072. 46. 81.	2.77 BRAND B4 1036. 24. 2.00	SUPERPLSTCZR BRAND B2 25.4 -0	.357 .249 201. 4.0 2.0	4.00 -0 104.0 60. MIN DAMP,73F	(28-DAY) •11210. • 10700. 11440. 11490.	(-0-DAY) -0 • -0 -0 -0	(28-DAY) -0 • -0 -0 -0	(28-DAY) -0 • -0 -0 -0	(28-DAY) -0. -0 -0 -0
CIBAB 3/ 9/83 5.9/2.0 (T)	BRAND C6 II 564. 11.	BRAND A2 CLASS C 242. 5. 30.	1/2 LIMESTONE BRAND E4 2069. 46. 91.	2.77 BRAND B4 1035. 24. 2.00	SUPERPLSTCZR BRAND B2 30.7 -0	.360 .252 203. 4.1 2.0	4.00 -0 102.0 90. MIN DAMP,73F	(28-DAY) •11430. • 11710. 10700. 11880.	(-0-DAY) -0 • -0 -0 -0	(28-DAY) -0 • -0 -0 -0	(28-DAY) -0 • -0 -0 -0	(28-DAY) -0. -0 -0 -0
CIB C 3/15/83 10.0/2.0 (Q)	BRAND C6 II 933. 18.	NONE 0 0 0	1/2 LIMESTONE BRAND E4 1785. 40. 78.	2.77 BRAND B4 893. 21. 2.00	RDUCR/RTRDER BRAND C1 4.5 -0	.364 .364 340. 4.1 2.0	2.75 -0 102.0 60. MIN DAMP,73F	(28-DAY) • 9470. • 9210. 9730. 9460.	(-0-DAY) -0 • -0 -0 -0	(28-DAY) -0 • -0 -0 -0	(28-DAY) -0 • -0 -0 -0	(28-DAY) -0. -0 -0 -0
CIB C 3/15/83 10.0/2.0 (Q)	BRAND C6 II 930. 18.	NONE 0 0 0	1/2 LIMESTONE BRAND E4 1780. 39. 69.	2.77 BRAND B4 890. 21. 2.00	RDUCR/RTRDER BRAND C1 4.5 -0	.369 .369 344. 4.2 2.0	3.00 -0 103.0 90. MIN DAMP,73F	(28-DAY) • 9050. • 8670. 8970. 9510.	(-0-DAY) -0 • -0 -0 -0	(28-DAY) -0 • -0 -0 -0	(28-DAY) -0 • -0 -0 -0	(28-DAY) -0. -0 -0 -0
CLB B 3/17/83 8.5/2.0 (R)	BRAND C7 II 803. 15.	NONE 0 0 0	1/2 LIMESTONE BRAND E4 2079. 46. 81.	2.77 BRAND B4 1040. 24. 2.00	SUPERPLSTCZR BRAND BD 14.8 RDUCR/RTRDER 5.0	.266 .266 214. 3.0 2.0	3.50 -0 104.0 60. MIN DAMP,73F	(28-DAY) •11490. • 12229. 10750. 0	(-0-DAY) -0 • -0 -0 -0	(28-DAY) -0 • -0 -0 -0	(28-DAY) -0 • -0 -0 -0	(28-DAY) -0. -0 -0 -0

MIX I.D. MIX DATE CF/CAFA	CEMENT BRAND TYPE LBS/CUYD PCT VOLUME	FLYASH BRAND CLASS LBS/CUYD PCT VOLUME PCT REPLACED	COARSE AGG SIZE MATERIAL SOURCE LBS/CUYD P:VOL,DRUW	FINE AGG FINENESS SOURCE LBS/CUYD PCT VOLUME CA/FA(LB/LB)	ADMIXTURE TYPE BRAND DOSE(OZ/100) (2ND TYPE) (2ND DOSE)	WATER W/C W/B LB/CUYD GAL/SACK PCT AIR	MISC SLUMP UNIT WT MIX TMP MX TIME CURING	TEST RESULTS 6 X 12 CYLINDER (STEEL) (PSI)	6 X 12 CYLINDER (STEEL) (PSI)	6X6X18 BEAM (STEEL) (PSI)	4 X 8 CYLINDER (CARDBD) (PSI)	4 X 8 CYLINDER (STEEL) (PSI)
C!B B	BRAND C7		1/2	2.77	SUPERPLSTCZR	.270	3.50	(28-DAY)	(-0-DAY)	(28-DAY)	(28-DAY)	(28-DAY)
3/17/83	II	NONE	LIMESTONE	BRAND B4	BRAND BD	.270	-0	•11820. •	-0 •	-0 •	-0 •	-0•
8.5/2.0	802.	0	BRAND E4	1038.	21.1	217.	99.0	11880.	-0	-0	-0	-0
(R)	15.	0	2075.	24.	RDUCR/RTRDER	3.0	90. MIN	11650.	-0	-0	-0	-0
		0	46. 91.	2.00	5.0	2.0	DAMP,73F	11940.	-0	-0	-0	-0
C!BAC	BRAND C7	BRAND A2	1/2	2.77	RDUCR/RTRDER	.427	2.75	(28-DAY)	(-0-DAY)	(28-DAY)	(28-DAY)	(28-DAY)
3/16/83	II	CLASS C	LIMESTONE	BRAND B4	BRAND C1	.299	-0	• 9650. •	-0 •	-0 •	-0 •	-0•
7.0/2.0	662.	283.	BRAND E4	923.	5.0	283.	101.0	10330.	-0	-0	-7	-0
	12.	6.	1848.	21.		4.8	60. MIN	9410.	-0	-0	-0	-0
(S)		30.	41. 72.	2.00	-0	2.0	DAMP,73F	9200.	-0	-0	-0	-0
C!BAC	BRAND C7	BRAND A2	1/2	2.77	RDUCR/RTRDER	.457	2.75	(28-DAY)	(-0-DAY)	(28-DAY)	(28-DAY)	(28-DAY)
3/16/83	II	CLASS C	LIMESTONE	BRAND B4	BRAND C1	.320	-0	• 9590. •	-0 •	-0 •	-0 •	-0•
7.0/2.0	654.	280.	BRAND E4	913.	5.0	299.	101.0	9850.	-0	-0	-0	-0
	12.	6.	1826.	21.		5.2	90. MIN	9250.	-0	-0	-0	-0
(S)		30.	40. 71.	2.00	-0	2.0	DAMP,73F	9660.	-0	-0	-0	-0
C!RAB	BRAND C7	BRAND A2	1/2	2.77	SUPERPLSTCZR	.348	4.00	(28-DAY)	(-0-DAY)	(29-DAY)	(28-DAY)	(28-DAY)
3/22/83	II	CLASS C	LIMESTONE	BRAND B4	BRAND BD	.244	-0	•12170. •	-0 •	-0 •	-0 •	-0•
5.9/2.0	566.	243.	BRAND E4	1039.	24.5	197.	105.0	12060.	-0	-0	-0	-0
	11.	6.	2078.	24.	RDUCR/RTRDER	3.9	60. MIN	12220.	-0	-0	-0	-0
(T)		30.	46. 81.	2.00	5.0	2.0	DAMP,73F	12220.	-0	-0	-0	-0
C!BAB	BRAND C7	BRAND A2	1/2	2.77	SUPERPLSTCZR	.355	4.50	(28-DAY)	(-0-DAY)	(29-DAY)	(28-DAY)	(28-DAY)
3/22/83	II	CLASS C	LIMESTONE	BRAND B4	BRAND BD	.249	-0	•12150. •	-0 •	-0 •	-0 •	-0•
5.?/2.0	565.	243.	BRAND E4	1036.	35.1	201.	105.0	11710.	-0	-0	-0	-0
	11.	5.	2073.	24.	RDUCR/RTRDER	4.0	90. MIN	12220.	-0	-0	-0	-0
(T)		30.	46. 81.	2.00	5.0	2.0	DAMP,73F	12560.	-0	-0	-0	-0
C!BAB	BRAND C7	BRAND A2	1/2	2.77	SUPERPLSTCZR	.360	5.00	(28-DAY)	(-0-DAY)	(28-DAY)	(28-DAY)	(28-DAY)
6/ 7/83	II	CLASS C	LIMESTONE	BRAND B4	BRAND BD	.251	-0	•11390. •	-0 •	-0 •	-0 •	-0•
5.9/2.0	558.	239.	BRAND E4	1042.	21.0	200.	98.0	11440.	-0	-0	-0	-0
	11.	5.	2083.	24.	RDUCR/RTRDER	4.0	60. MIN	11510.	-0	-0	-0	-0
(T)		30.	46. 81.	2.00	2.0	2.0	DAMP,73F	11230.	-0	-0	-0	-0

ADDITIONAL DATA FROM MIX ABOVE: 1) 6X12 COMPR STEEL 1D • 4690.• 4560. 4790. 4720.
2) 6X12 COMPR STEEL 7D • 9460.• 9440. 9740. 9200.

	CEMENT	FLYASH	COARSE AGG	FINE AGG	ADMIXTURE	WATER	MISC	••••••••••••••• TEST RESULTS				•••••••••••••••
I.D. DATE /CAFA	BRAND TYPE LBS/CUYD PCT VOLUME	BRAND CLASS LBS/CUYD PCT VOLUME PCT REPLACED	SIZE MATERIAL SOURCE LBS/CUYD P:VOL,DRUM	FINENESS SOURCE LBS/CUYD PCT VOLUME CA/FA(LB/LB)	TYPE BRAND DOSE(OZ/100) (2ND TYPE) (2ND DOSE)	W/C W/B LB/CUYD GAL/SACK PCT AIR	SLUMP UNIT WT MIX TMP MX TIME CURING	6 X 12 CYLINDER (STEEL) (PSI)	6 X 12 CYLINDER (STEEL) (PSI)	6X6X18 BEAM (STEEL) (PSI)	4 X 8 CYLINDER (CARDBD) (PSI)	4 X 8 CYLINDER (STEEL) (PSI)
CEBAB 6/ 8/83 5.5/2.0 (T)	BRAND C7 II 564. 11.	BRAND A2 CLASS C 242. 5. 30.	1/2 LIMESTONE BRAND E4 2109. 47. 92.	2.77 BRAND B4 1055. 24. 2.80	SUPERPLSTCZR BRAND B0 25.1 RDUCR/RTRDER 4.0	.322 .226 182. 3.6 2.0	5.00 -0 • 78.0 15. MIN DAMP,73F	(-0-DAY) -0 • -0 -0 -0	(-0-DAY) -0 • -0 -0 -0	(-0-DAY) -0 • -0 -0 -0	(-0-DAY) -0 • -0 -0 -0	(-0-DAY) -0• -0 -0 -0

ADDITIONAL DATA FROM MIX ABOVE:

1M	4X8 VIBRTD, MOIST 73	•10170.•	8040.	9790.	10540.
2M	4X8 MOIST 14, DRY 14	•11480.•	11280.	10550.	12610.
3M	4X8 MOIST 7, DRY 21	•11380.•	9110.	11050.	11710.
4M	4X8 M 14, HOT+DRY 14	•12360.•	9750.	12730.	11980.
5M	4X8 M 7, HOT+DRY 21	•12260.•	12140.	11900.	12730.
6M	4X8 UNDER WATER 28	•11050.•	10150.	7720.	11940.
7M	4X8 HIGH STRNGTH CAP	•10550.•	10350.	10740.	7760.
8M	4X8 MISC. CAP	•11180.•	10420.	11380.	11740.

Bibliography

1. Albinger, John, and Moreno, Jaime, "High-Strength Concrete, Chicago Style," Concrete Construction, Vol. 26, No. 3, March 1981, pp. 241-245.

2. Alexander, K. M., Bruere, G. M., and Ivanusec, I., "The Creep and Related Properties of Very High-Strength Superplasticized Concrete," Cement and Concrete Research Journal, Vol. 10, No. 2, March 1980, pp. 131-137.

3. Anderson, Arthur R., "Research Answers Needed for Greater Utilization of High Strength Concrete," Journal of the Prestressed Concrete Institute, Vol. 25, No. 4, July-August 1980, pp. 162-164.

4. Berry, E. E., and Malhotra, V. M., "Fly Ash for Use in Concrete--A Critical Review," Journal of the American Concrete Institute, Proceedings, Vol. 77, No. 2, March-April 1980, pp. 59-73.

5. Bickley, J. A., and Payne, J. C., "High Strength Cast in Place Concrete in Major Structures in Ontario," Symposium on Practical and Potential Applications of High-Strength Concrete, American Concrete Institute Annual Convention, March 22, 1979, Milwaukee, Wisconsin, 27pp.

6. Bickley, J. A., "Concrete Optimization," Concrete International, Vol. 4, No. 6, June 1982, pp. 38-41.

7. Blick, Ronald L., "Some Factors Influencing High-Strength Concrete," Modern Concrete, April 1973, pp. 38-41, 47.

8. Bloem, Delmar L., and Gaynor, Richard D., "Effects of Aggregate Properties on Strength of Concrete," Journal of the American Concrete Institute, Proceedings, Vol. 60, No. 10, October 1963, pp. 1435-1453.

9. Brooks, J. J., Wainwright, P. J., and Neville, A. M., "Time-Dependent Behavior of High-Early-Strength Concrete Containing a Superplasticizer," SP 68-5, Developments in the Use of Superplasticizers, American Concrete Institute, Detroit 1981, pp. 81-100.

10. Burgess, A. J., Ryell, J., and Bunting, J., "High Strength Concrete for the Willows Bridge," Journal of the American Concrete Institute, Proceedings, Vol. 67, No. 8, August 1970, pp. 611-619.

11. Carino, Nicholas J., and Lew, H. S., "Re-examination of the Relation between Splitting Tensile and Compressive Strength of Normal Weight Concrete," Journal of the American Concrete Institute, Proceedings, Vol. 79, No. 3, May-June 1982, pp. 214-219.

12. Carpenter, James E., "Applications of High Strength Concrete for Highway Bridges," Public Roads, Vol. 44, No. 2, Sept. 1980, pp. 76-83.

13. Carrasquillo, R. L., Nilson, A. H., and Slate, F. O., "Properties of High-Strength Concrete Subject to Short-Term Loads," Journal of the American Concrete Institute, Proceedings, Vol. 78, No. 3, May-June 1981, pp. 171-178.

14. Carrasquillo, R. L., Nilson, A. H., and Slate, F. O., "The Production of High-Strength Concrete," Department Report No. 78-1, Structural Engineering Department, Cornell University, Ithaca, New York, May 1978, 91pp. Also, "The Production of High-Strength Concrete," by Ramon L. Carrasquillo, MSc Thesis, Cornell University, Ithaca, New York, May 1978, 90pp.

15. Carrasquillo, R. L., Slate, F. O., and Nilson, A. H., "Microcracking and Behavior of High-Strength Concrete Subject to Short-Term Loads," Journal of the American Concrete Institute, Proceedings, Vol. 78, No. 3, May-June 1981, pp. 179-186.

16. Chicago Committee on High-Rise Buildings, High Strength Concrete in Chicago High-Rise Buildings, Task Force Report No. 5, February 1977, 63pp.

17. Colaco, J. P., Ames, J. B., and Dubinsky, E., "Concrete Shear Walls and Spandrel Beam Moment Frame Brace New York Office Tower," Concrete International, Vol. 3, No. 6, June 1981, pp. 23-28.

18. Cook, James E., "Research and Application of High-Strength Concrete Using Class C Fly Ash," Concrete International, Vol. 4, No. 7, July 1982, pp. 72-80.

19. Cordon, William A., and Gillespie, H. Aldridge, "Variables in Concrete Aggregates and Portland Cement Paste Which Influence the Strength of Concrete," Journal of the American Concrete Institute, Proceedings, Vol. 60, No. 8, August 1963, pp. 1029-1049.

20. Day, Ken W., "Quality Control of 55 MPa Concrete for Collins Place Project, Melbourne, Australia," Concrete International, Vol. 3, No. 3, March 1981, pp. 17-24.

21. Dezhen, G., Dayu, X., and Zhang, L., "Model of Mechanism for Naphthalene Series Water-Reducing Agent," Journal of the American Concrete Institute, Proceedings, Vol. 79, No. 5, September-October 1982, pp. 378-386.

22. "Fly Ash in Concrete, Part I," Concrete Construction, Vol. 27, No. 5, May 1982, pp.417-421.

23. "Fly Ash in Concrete, Part II," Concrete Construction, Vol. 27, No. 5, May 1982, pp. 424-427.

24. Freedman, Sidney, "High-Strength Concrete," Publication No. 1S176, Portland Cement Association, 1971, 19pp. (Reprint from Modern Concrete, October, November, December 1970, and January, February 1971.)

25. French, P. J., Montgomery, R. G. J., and Robson, T. D., "High Concrete Strength Within the Hour," Concrete, Vol. 5, No. 8, August 1971, pp. 253-258.

26. Gaynor, R. D., "An Outline on High Strength Concrete," Paper orally presented at the Annual Convention of Virginia Ready Mixed Concrete Association, April 5, 1973, 14pp.

27. Harris, A. J., "High-Strength Concrete: Manufacture and Properties," The Structural Engineer, Vol. 47, No. 11, November 1969, pp. 441-446.

28. Harris, A. J., "Ultra High Strength Concrete," Journal of the Prestressed Concrete Institute, Vol. 12, No. 1, February 1967, pp. 53-59.

29. Hattori, Kenichi, "Experiences with Mighty Superplasticizers in Japan," SP 62-3, Superplasticizers in Concrete, American Concrete Institute, Detroit 1979, pp. 37-66.

30. Hester, Weston T., "Field-Testing High-Strength Concretes: A Critical Review of the State-of-the-Art," Concrete International, Vol. 2, No. 12, December 1980, pp. 27-38.

31. Hester, Weston, T., "Superplasticizers: Lessons from Three Tough Jobs," A Presentation for the World of Concrete, November 15-16, 1982, 16pp.

32. Hewlett, P. C., "The Concept of Superplasticized Concrete," SP 62-1, Superplasticizers in Concrete, American Concrete Institute, Detroit 1979, pp. 1-20.

33. "High-Strength Concrete," Building (London), Vol. 211, No. 6436, 1966, pp. 129-130.

34. "High Strength Concrete," Concrete, February 1970, pp. 83-84.

35. "High Strength Concrete," First Edition, National Crushed Stone Association January 1975, 16pp.

36. "High-Strength Concrete . . . Crushed Stone Aggregate Makes the Difference," Presented at the January 1975 National Crushed Stone Association in Florida, November 1974, 31pp.

37. Hognestad, E., and Perenchio, W. F., "Developments in High-Strength Concrete," Proceedings of the Seventh Congress of the Federation Internationale de la Precontrainte, Vol. 2 - Lectural and General Reports, New York, May 26-June 1, 1974, pp. 21-24.

38. Holbek, Kai, and Skrastins, Janis I., "Some Experience with the Use of Superplasticizers in the Precast Concrete Industry in Canada," SP 62-5, Superplasticizers in Concrete, American Concrete Institute, Detroit, 1979, pp. 123-136.

39. Hollister, S. C., "Urgent Need for Research in High-Strength Concrete," Journal of the American Concrete Institute, Proceedings, Vol. 73, No. 3, March 1976, pp. 136-137.

40. "How Super are Superplasticizers?", Concrete Construction, Vol. 27, No. 5, May 1982, pp. 409-415.

41. Johnston, Colin D., "Fifty-Year Developments in High Strength Concrete," Journal of the Construction Division, American Society of Civil Engineers, December 1975, pp. 801-818.

42. Jones, R., and Kaplan, M. F., "The Effect of Coarse Aggregate on the Mode of Failure of Concrete in Compression and Flexure," Magazine of Concrete Research, August 1957, pp. 89-94.

43. Kaar, P. H., Hanson, N. W., and Capell, H. T., "Stress-Strain Characteristics of High-Strength Concrete," Research and Development Bulletin RD 051.01D, Portland Cement Association, 1977, 11pp.; also, Publication SP-55, Douglas McHenry International Symposium on Concrete and Concrete Structures, American Concrete Institute, Detroit, Mich., 1978, pp. 161-185.

44. Kaplan, M. F., "Flexural and Compressive Strength of Concrete as Affected by the Properties of Coarse Aggregates," Journal of the American Concrete Institute, Proceedings, Vol. 56, No. 5, May 1959, pp. 1193-1208.

45. Kaplan, M. F., "Ultrasonic Pulse Velocity, Dynamic Modulus of Elasticity, Poisson's Ratio and the Strength of Concrete Made with Thirteen Different Coarse Aggregates," Bulletin Rilem No. 1, The International Union of Testing and Research Laboratories for Materials and Structures, March 1959, pp. 59-73.

46. Kirsten, Eriken, and Nepper-Christensen, Palle, "Experiences in the Use of Superplasticizers in Some Special Fly Ash Concretes," Sp 68-1, Developments in the Use of Superplasticizers, American Concrete Institute, Detroit, 1981, pp. 1-20.

47. Lane, R. O., and Best, J. F., "Properties and Use of Fly Ash in Portland Cement Concrete," Concrete International, Vol. 4, No. 7, July 1982, pp. 81-92.

48. Leslie, K. E., Rajagopalan, K. S., and Everard, N. J., "Flexural Behavior of High-Strength Concrete Beams," Journal of the American Concrete Institute, Proceedings, Vol. 73, No. 9, September 1976, pp. 517-521. (Based on "Flexural Behavior of Rectangular Beams of Extra High Strength Concrete," by Keith E. Leslie, unpublished MSc Thesis, The University of Texas at Arlington, 1975, 101pp.)

49. Macinnis, C., and Kosteniuk, P. W., "Effectiveness of Revibration and High-Speed Slurry Mixing for Producing High-Strength Concrete," Journal of the American Concrete Institute, Proceedings, Vol. 76, No. 12, December 1979, pp. 1255-1265.

50. Macinnis, Cameron, and Thomson, Donald V., "Special Techniques for Producing High Strength Concrete," Journal of the American Concrete Institute, Proceedings, Vol. 67, No. 12, December 1970, pp. 996-1002.

51. McKerral, William C., Ledbetter, W. B., and Teague, D. J., "Analysis of Fly Ashes Produced in Texas," Report No. FHWA/TX-81/21 + 240-1, Texas Transportation Institute, The Texas A & M University System, Texas State Department of Highways and Public Transportation, January 1981, 97pp.

52. Malhotra, V. M., "Development of Sulfur-Infiltrated High-Strength Concrete," Journal of the American Concrete Institute, Proceedings, Vol. 72, No. 9, September 1975, pp. 466-473.

53. Malhotra, V. M., "Mechanical Properties and Durability of Superplasticized Semi-Lightweight Concrete," SP 68-16, Developments in the Use of Superplasticizers, American Concrete Institute, Detroit 1981, pp. 283-306.

54. Manual of Testing Procedures, Volume 2, 400-A Series, Texas State Department of Highways and Public Transportation, Revised January 1, 1978.

55. Mather, Bryant, "High-Compressive-Strength Concrete, A Review of the State of the Art," Technical Documentary Report No. AFSWC-TDR-62-56, Air Force Special Weapons Center, Kirtland AF Base, New Mexico, August 1962, 90pp.

56. Mather, Katherine, "High Strength, High Density Concrete," Journal of the American Concrete Institute, Proceedings., Vol. 62, August 1965, pp. 951-962.

57. Meininger, Richard C., "Use of Fly Ash in Cement and Concrete--Report of Two Recent Meetings," Concrete International, Vol. 4, No. 7, July 1982, pp. 52-57.

58. "Methods of Achieving High Strength Concrete," Journal of the American Concrete Institute, Proceedings, Vol. 64, No. 1, January 1967, pp. 45-48.

59. Morgan, Austin, H., "High-Strength Ready-Mixed Concrete," Paper presented to 41st Annual Convention of National Ready Mixed Concrete Association, January 1971, 18pp.

60. Murata, J., Kawai, T., and Kokubu, K., "Studies on the Utilization of Water-Reduced High-Strength Concrete Piers," SP 68-3, Developments in the Use of Superplasticizers, American Concrete Institute, Detroit, 1981, pp. 41-60.

61. Nagataki, Shigeyoshi, and Yonekura, Asuo, "Studies of the Volume Changes of High Strength Concretes with Superplasticizer," Journal of the Japan Prestressed Concrete Engineering Association, Vol. 20, 1978, pp. 26-33.

62. Nagataki, Shigeyoshi, "On the Use of Superplasticizers," Journal of the Japan Prestressed Concrete Engineering Association, Vol. 20, 1978, pp. 7-15.

63. "New York City Gets its First High-Strength Concrete Tower," Engineering News-Record, Vol. 201, No. 18, November 2, 1978, p. 22.

64. Ngab, A. S., Nilson, A. H., and Slate, F. O., "Shrinkage and Creep of High-Strength Concrete," Journal of the American Concrete Institute, Proceedings, Vol. 78, No. 4, July-August 1981, pp. 255-261.

65. Nilson, Arthur H., and Slate, Floyd O., "Structural Properties of Very High Strength Concrete," Second Progress Report, Department of Structural Engineering, School of Civil and Environmental Engineering, Cornell University, January 1979, 62pp.

66. 1980 Annual Book of ASTM Standards, Part 14, Concrete and Mineral Aggregates, American Society for Testing and Materials, Philadelphia.

67. 1982 Standard Specifications for Construction of Highways, Streets and Bridges, Texas State Department of Highways and Public Transportation, 900pp.

68. Okada, Kiyoshi, and Azimi, M. Azam, "Strength and Ductility of Reinforced High Strength Concrete Beams," Memoirs, Faculty of Engineering, Kyoto University, Vol. 43, Part 2, April 1981, pp. 304-318.

69. Parrot, L. J., "The Production and Properties of High-Strength Concrete," Concrete, November 1969, pp. 443-448.

70. Parrot, L. J., "The Properties of High-Strength Concrete," Technical Report No. 417, May 1969, Cement and Concrete Association, London, 12pp.

71. Parrot, L. J., "The Selection of Constituents and Proportions for Producing Workable Concrete with a Compressive Cube Strength of 80 to 110 N/mm^2 (11,600 to 15,900 lbf/in^2)," Technical Report No. 416, Cement and Concrete Asssociation, May 1969, 12pp.

72. Perenchio, W. F., "An Evaluation of Some of the Factors Involved in Producing Very High-Strength Concrete," Bulletin No. RD014, Portland Cement Association, 1973, 7pp.

73. Pfeiffenberger, Lucas E., and Ray, Thomas B., "Use of a Superplasticizer in the Manufacture of Extra-High-Strength Block," SP 68-2, Developments in the Use of Superplasticizers, American Concrete Institute, Detroit, 1981, pp. 21-40.

74. Popovics, Sandor, "Strength Relationships for Fly Ash Concrete," Journal of the American Concrete Institute, Proceedings, Vol. 79, No. 1, January-February 1982, pp. 43-49.

75. Roy, D. M., Gouda, G. R., and Bobrowsky, A., "Very High Strength Cement Pastes Prepared by Hot Pressing and Other High Pressure Techniques," Cement and Concrete Research, Vol. 2, No. 3, May 1972, pp. 349-366.

76. Roy, D. M., and Gouda, G. R., "High Strength Generation in Cement Pastes," Cement and Concrete Research, Vol. 3, No. 6, November 1973, pp. 807-820.

77. Russell, H. G., and Corley, W. G., "Time-Dependent Behavior of Columns in Water Tower Place," Research and Development Bulletin RD052.01B, "Portland Cement Association, 1977, 10pp. Also, Douglas McHenry International Symposium on Concrete and Concrete Structures, SP-55, American Concrete Institute, Detroit, Michigan, 1978, pp. 347-374.

78. Ryan, David J., "A First Hand Report of High Strength Concrete in Houston, Texas," Southwestern Laboratories, Inc., Houston, Texas 1980, 41pp.

79. Ryell, John, "High Strength Concrete," Part I, Canadian Pit and Quarry, January 1970, pp. 16-19.

80. Ryell, John, "High Strength Concrete," Part II, Canadian Pit and Quarry, February 1970, pp. 26-28.

81. Saucier, K. L., "Determination of Practical Ultimate Strength of Concrete", Miscellaneous Paper C-72-16, U.S. Army Engineer Waterways Experiment Station, Vicksburg, Mississippi, June 1972, 29pp.

82. Saucier, Kenneth L., "High-Strength Concrete, Past, Present, Future," Concrete International, Vol, 2, No. 6, June 1980, pp. 46-50.

83. Saucier, K. L., Tynes, Wm. O., and Smith, E. F., "High-Compressive-Strength Concrete," Technical Report No. AFWL-TR-65-16, Air Force Weapons Laboratory, Kirtland Air Force Base, New Mexico, September 1965, 83pp.

84. Shah, S. P. (Ed.), Proceedings of the Workshop on High-Strength Concrete, University of Illinois at Chicago Circle, Chicago, Illinois, sponsored by the National Science Foundation, December 2-4, 1979, 226pp.

85. Shah, S. P., "High-Strength Concrete--A Workshop Summary," Concrete International, Vol. 3, No. 5, May 1981, pp. 94-98.

86. Shilstone, James M., "Concrete Strength Loss and Slump Loss in Summer," Concrete Construction, Vol. 27, No. 5, May 1982, pp. 429-432.

87. Simmons, D. D., Pasko, T. J., Jr., and Jones, W. R., "Properties of Portland Cement Concretes Containing Pozzolanic Admixtures," Report No. FHWA/RD-80-184, Federal Highway Administration, U.S. Dept. of Transportation, April 1981, 45pp.

88. Smith, E. F., Tynes, Wm. O., and Saucier, K. L., "High-Compressive-Strength Concrete, Development of Concrete Mixtures," Technical Documentary Report No. RTD TDR-63-3114, U.S. Army Engineer Experiment Station, Vicksburg, Mississippi, February 1964, 45pp.

89. Standard Specifications for Highways Bridges, American Association of State Highway and Transportation Officials, 12th ed., Washington, D. C., 1977 (also 1978-1982 Interim Provisions).

90. State-of-the-Art Report, High Strength Concrete, Fourth Draft, ACI Committee 363, American Concrete Institute, Detroit, 1983.

91. Stein, Jane, "A Pillar of Strength," Mosaic, April 1982, pp. 23-26.

92. "Structural Trends in New York City Buildings," Civil Engineering, ASCE, Vol. 53, No. 1, January 1983, pp. 30-37.

93. "Superplasticizing Admixtures in Concrete," Report of a Joint Working Party of the Cement Admixtures Association and the Cement and Concrete Association, June 1978, 32pp.

94. "The World's Tallest Concrete Buildings--Today and Yesterday," Concrete Construction, Vol. 28, No. 2, February 1983, pp. 91-100.

95. Tognon, Giampietro, and Cangiano, Stefano, "Air Contained in Superplasticized Concretes," Journal of the American Concrete Institute, Proceedings, Vol. 79, No. 5, September-October 1982, pp. 350-354.

96. Tognon, G., Ursella, P., and Coppetti, G., "Design and Properties of Concretes with Strength over 1500 kgf/cm^2," Journal of the American Concrete Institute, Proceedings, Vol. 77, No. 3, May-June 1980, pp. 171-178.

97. "Tower Touches Few Bases," Engineering News-Record, June 16, 1983, pp. 23-25.

98. Valore, R. C., Kudrenski, W., and Gray, D. E., "Application of High-Range Water-Reducing Admixtures in Steam-Cured Cement-Fly Ash Concretes," SP 62-17, Superplasticizers in Concrete, American Concrete Institute, Detroit, 1979, pp. 337-374.

99. Vivesvaraya, H. C., Desayi, P., and Babu, Shri K. H., "High Strength Concrete Mix Design, A Case Study," Special Publication SP-2, Cement Research Institute of India, March 1970, 28pp.

100. Vogt, Woodward L., Bernacki, F. W., and Lyles, Arthur T., "An Engineering Laboratory's Experience with Fly Ash," unpublished report, 31pp.

101. Walker, Stanton, and Bleom, Delmar L., "Effects of Aggregate Size on Properties of Concrete," Journal of the American Concrete Institute, Proceedings, Vol. 57, No. 9, September 1960, pp. 283-297.

102. Walz, Kurt, "The Production of High Strength Concrete," The Cement Marketing Company Limited, London, June 1966, p. 7, Translation Tec. 2037/R39.

103. Wang, P. T., Shah, S. P., and Naaman, A. E., "High-Strength Concrete in Ultimate Strength Design," Journal of the Structural Division, American Society of Civil Engineers, November 1978, pp. 1761-1773.

104. Woolgar, G., and Oates, D. B., "Fly Ash and the Ready-Mixed Concrete Producer," Concrete International, Vol. 1, No. 11, November 1979, pp. 34-40.

105. Yamamoto, Y., and Kobayashi, M., "Use of Mineral Fines in High Strength Concrete--Water Requirement and Strength," Concrete International, Vol. 4, No. 7, July 1982, pp. 33-40.

106. Zaitsev, Y. B., and Wittman, F. H., "Simulation of Crack Propagation and Failure of Concrete," Materials and Structures, Research and Testing (Paris), Vol. 14, No. 83, September-October 1981, pp. 357-365.

107. Zia, Paul, "Structural Design with High-Strength Concrete," Report No. PZIA-77-01, Department of Civil Engineering, North Carolina State University, March 1977, 65pp.